UNAPOLOGETIC
FREEDOM

UNAPOLOGETIC FREEDOM

HOW BITCOIN DEFEATS CENSORSHIP, ENSURES SOVEREIGNTY, AND RECLAIMS OUR LIBERTY FOREVER

JUSTIN REZVANI
FOREWORD BY JP SEARS

A ZION PUBLICATION

Copyright © 2022 Justin Rezvani

All rights reserved. No part of this publication may be reproduced, distributed, or transmitted in any form or by any means, including photocopying, recording, or other electronic or mechanical methods, without prior written permission of the author, except in the case of brief quotations embodied in critical reviews and certain other noncommercial uses permitted by copyright law.

No Investment Advice Disclaimer: The content in this book is for informational purposes only. You should not construe any such information or other material as legal, tax, investment, financial, or other advice. Nothing contained in this book constitutes a solicitation, recommendation, endorsement, or offer by Justin Rezvani, JP Sears, Modern Foundry Inc., or any third-party service provider to buy or sell any securities or other financial instruments in this or in any other jurisdiction in which such solicitation or offer would be unlawful under the securities laws of such jurisdiction. All content in this book is information of a general nature and does not address the circumstances of any particular individual or entity. Nothing in this book constitutes professional and/or financial advice, nor does any information in this book constitute a comprehensive or complete statement of the matters discussed or the law relating thereto. You alone assume the sole responsibility of evaluating the merits and risks associated with the use of any information or other content in this book before making any decisions based on such information or other content.

Paperback ISBN: 979-8-9856463-1-3
Hardcover ISBN: 979-8-9856463-0-6
E-Book ISBN: 979-8-9856463-3-7
Audio Book ISBN: 979-8-9856463-2-0

Independently published.
First printing, 2022
Book cover and layout design by *the*BookDesigners
www.getzion.com

CONTENTS

Foreword vii
Introduction 1

CHAPTER 1
The Current State of Censorship 15

CHAPTER 2
The Centralized vs. Decentralized Centuries .. 27

CHAPTER 3
Platform vs. Peer Governance............. 47

CHAPTER 4
Creating a More Civil, Secure, and
Accountable Online Discourse 69

CHAPTER 5
Building a Better Online Experience
Using Bitcoin and the Lightning Network.... 85

CHAPTER 6
Unapologetic Freedom.................. 103

Notes 116
Acknowledgements 119
About the Author...................... 121
End Notes 125

FOREWORD

As Winston Churchill, that stout defender of freedom and liberty, once said, these are "anxious and baffling times." We live in an age when Big Tech, Big Pharma, and Big Government have teamed up to make our rights to freedom of speech and expression smaller, while supersizing their own control and profiting from dividing, muzzling, and restricting us. Even as you read these words, our freedom is being eroded away by censorship, exploitative algorithms, and deplatforming. Which is why this book is so important.

Until a year or so ago, I thought Bitcoin was silly and created a video making fun of it as though it was just a giant pyramid scheme (full disclosure: because that's what I thought it was). But in the months since, I've realized that Bitcoin the asset and the network isn't some passing fad, but rather the digital infrastructure that we've been needing in order to reclaim our sovereignty. I also discovered that by solely relying on traditional social media platforms, we're consenting to Big Tech censoring, stealing, and selling our data, manipulating our minds, and strip mining our attention for the good of their advertisers (which makes it very expensive for us).

Many freedom lovers would agree with me that this is not OK. And now we understand that we can put an end to it – we just need to know how. For all the freedom-restricting problems that are caused by Big Tech, I used to think, "We need a solution, but unfortunately, we don't have one." That's not the case anymore, as I've learned that Bitcoin offers the solution that we need.

My self-education on these topics was a stark reminder that none of us can know what we don't know. I once heard Ram Dass paraphrasing Russian philosopher George Gurdjieff as saying, "If you're going to escape from prison, the first thing you must appreciate is that you are in one." My hope is that once you finish reading this book, you'll know that you are in that prison. And armed with this knowledge, you'll make the conscious choice to wake up, break free from jail, and boldly stride out of the old, restrictive paradigm into the new, expansive one. In the coming pages, Justin will demonstrate just how seriously our democratic way of life is being threatened and will then provide a solution that you can start using today.

I'm writing these words to you at a pivotal time. I am passionate about this topic because my livelihood has been under threat. There's a realistic fear that I could be deplatformed from all the centralized social media and video services at any moment (and quite honestly, at times, I'm surprised I haven't been). If that happens, it's not just my career as a comedian that's in jeopardy but also my ability to provide for my wife and child and connect to the millions of people around the world that I'm privileged enough to call friends and fans.

Thousands of fellow creators have already been kicked out of the online town hall because those who hold the keys decided that by daring to share independent opinions, writers, filmmakers, and other artists were threatening the narrative that they try to control. So they were banned. Who's next to be ejected from the public discourse? It could be me, or it could be you.

That is, unless we decide to take back our power – in the form of digital property rights – to own our content and data, the freedom to say whatever we like to whoever we please, and the ability to do so in online communities that are free of censorship, suppression, and exploitation. We've allowed ourselves to be the product in the Big Tech attention economy for far too long. In our hearts, freedom lovers and sovereign individuals know it's time to take a strong stand, say "Enough!" and contribute to building something better.

Our humanity is based on the ability to sit down face-to-face with each other and have honest, candid conversations. To agree about some things and disagree about others. To look someone from a different background in the eye and find common ground with them. I believe this should be just as pure digitally as it is in person, which is why we're being called to evolve beyond the old, traditional, central systems that exploit, manipulate, and enslave us and move forward to decentralized, encrypted utilities that bring the humanity back to our online interaction.

FOREWORD

The countless hours we spend perusing centralized platforms every day perpetuate a business model built on a complete lack of transparency and, dare I even say, deception. The tech titans claim that their goal is connecting us for free, but in reality, it's extracting and monetizing our attention and pushing the agenda of powerful corporate and political actors. They continually consolidate their power with censorship, banning, and bias masquerading as independent "fact checking" (even though while I was writing this foreword, Facebook admitted in court that their fact checkers' "fact checks" were based on opinions, NOT facts).

Wouldn't you rather use a utility that's free of such trickery and overreach? One where there's honesty, transparency, and true freedom of expression? We've been in need of this for a long time but without an answer. Fortunately, we now have the opportunity to stop being of service to the desires of the Facebooks, Twitters, and Googles of the world and start using utilities that serve us instead.

The time we live in right now is pivotal enough that we don't have the luxury of sitting on the sidelines and hoping that someone else will bring about the change that we wish to see. In fact, the choice to remain spectators is now gone because life has ripped the bleachers out of the gladiator arena. We're all on the playing field now. As such, we only have three levels of a warriorship open to us:

1) Level one involves standing amidst the fray but not doing a damn thing as our free speech and liberties are eroded away because we are unaware of what's at stake.

2) At level two, we become aware of the change that is needed and what we have to do to make it happen, but we are too afraid to take action. So we sit back and still do nothing.

3) And at level three, we become true warriors who know exactly what we must do, muster the courage to take up our swords, and act decisively to help bring about the change we wish to see in the world.

As our battle for freedom is being fought in the digital Colosseum, we can begin winning the fight immediately by supporting decentralized companies who prioritize values above profits, voting with our dollars and attention, and bravely defending the principles we hold dear. In any given moment, we can choose to live out of fear and therefore avoid change, while allowing ourselves to go even deeper into the direction that the old paradigm of Big Tech wants to take us. Or we can choose freedom and better ways of doing things that empower us to drive ourselves forward to where we want to go. Live out of fear or freedom – the choice is ours.

This is bigger than you and me. It's also about the legacy we leave for our children and grandchildren. In this respect, there are only two possibilities on the table. If we continue to remain passive pawns to how centralized tech companies want us to be, we'll have to eventually explain to our grandkids what freedom was. Because if we don't force a change by then, censorship, control, and authoritarianism will have made freedom a memory rather than a reality. Or we can rise up now, take our stand, and proudly experience the joy of getting to tell our grandkids about what censorship was. Which sounds better to you?

Before you decide, I'd like to say thanks for picking up this book. It shows that you want to learn more about the predicament we find ourselves in so that you can not only take your power back but also contribute to the building of a better way that benefits other people. As I mentioned earlier, I used to know we needed to find a better path with social technology but didn't know what it was. That all changed when I met Justin, who's subsequently become a great friend and business partner.

Personally, I am an action taker, but before meeting Justin, I didn't know what to do in this area. Thanks to his mind-boggling depth of knowledge in the tech space, he opened my eyes to see how Bitcoin-based technology is actually the digital infrastructure of freedom and sovereignty. In other words, it's the solution to the Big Tech challenges we face. What Justin has gifted me is the peace of mind of knowing that there's a solution to a massive problem. Now I'm taking action in a meaningful way to give myself the empowered change I want and sharing it with the world around me.

Without Justin's visionary mind, I'd still be a frustrated man without answers. So it is an absolute pleasure to send you off into the rest of this book because you're about to be equipped with the knowledge to fill your life and future with more freedom by taking your power back.

If you and I don't take action right now, our world is in real trouble. It needs changemakers to lend a big helping hand. Getting back on our feet always starts with people who care, and you're one of them. Knowing there are many freedom lovers just like you and me, together we can summon the courage, conviction, and commitment to rise up out of the old swamp lands that don't serve us anymore. It's my honor to stand alongside you on this mission of reclaiming freedom in our daily and digital lives, because freedom is the very thing that makes life such a precious gift.

JP SEARS
Austin, Texas
January 2022

INTRODUCTION
WE ARE NOT AFFRAID

> "The further a society drifts from the truth, the more it will hate those who speak it."
>
> **SELWYN DUKE**

I hope this book found you without too much trouble and that these words haven't already been banned by online retailers and the very bookstores in which freedom of thought and expression should flourish. But if you had hassle obtaining this copy, you've already experienced the censorship, deplatforming, and other ill effects of centralization that motivated me to write this book.

If it only took you one click to get here, let's consider some other intrusions on your digital liberty that you're probably all too familiar with. Have you ever searched for someone on social media, only to find that their account had been deleted? Or maybe you've shared a post that a platform pop-up labeled as "misinformation." When you logged back into your feed later, it had mysteriously vanished.

If you don't use (and aren't used by) such platforms, then it's still likely that you've Googled something recently. In which case, you've been bombarded with just as many ads as you'd see on "the socials," and every result has been ordered, filtered, and scrubbed of the 10,000 or so websites

INTRODUCTION

that are blacklisted daily. Their supposed crime? Deviating from the mainstream, left-leaning narrative that Big Tech's advertisers want to see their product placements appear next to. And make no mistake about it, such search and [anti] social services might be marketed as free, but they exact a heavy toll on your mental health, attention, and energy in return.

Like Andrew Lewis tweeted at the beginning of the decentralized age ushered in by the release of the iPhone in 2010, "You're the product being sold" in this digital economy. The scope, scale, and reach of advertising and digital censorship has been amplified since the golden days of TV-based marketing by the proliferation of ubiquitous search engines, portable devices that deliver targeted ads 24/7/365, and the self-appointed kings of sovereign digital fiefdoms who wield more power than any Roman emperor.

Connecting Creators and Their Communities

Before we go any further with the big picture, allow me to share a little of my own story and why I became so passionate about defeating censorship, profit-driven centralization, and digital disenfranchisement. I graduated from Cal Poly Pomona when I was 22 and started working for an advertising agency. A 2013 conversation over dinner with my high school friend and actor Keegan Allen (star of the hit TV show *Pretty Little Liars*) proved to be pivotal. Seeking to learn more about how he interacted with fans, friends, and family members, I asked him, "Is there anything in the App Store that connects an influencer to a brand?"

He said no, but he wished there was, as that was exactly what he needed to come closer to his audience without a third-party arbiter in the middle. So I saw an opportunity to create an app called Reach (through the company theAmplify) that did what Keegan and other creators wanted. Part of my vision was to enable instant and direct payments for branded micro content on Instagram, without

INTRODUCTION

creators waiting for a wire transfer or check to clear. Lacking the benefit of cryptocurrency at the time, my team and I accomplished that through PayPal. The overwhelmingly positive response we got from the creative community sowed the seeds for what would eventually become Zion.

I ended up selling theAmplify in mid-2016. Based on conversations with Keegan and others, I knew that while our app solved some of the problems creators found in direct-to-brand monetization, there must be a way to go beyond advertising and create a connection between artists and their audiences. This notion didn't crystallize until the epiphany I had during an orthodontist appointment on November 30, 2019. I thought it was going to be a routine visit – in and out in 30 minutes, tops. But as soon as my orthodontist shined the bright light over my face, I blacked out and went into the light.

In my dreamlike state, I started following a happier and more fulfilled version of myself. After what felt like hours, I reentered my body through the same light that had taken me away. I was still sitting in my orthodontist's chair, but it was a different me.

In a cold sweat, I felt the color drain from my face and started feeling as if I was about to vomit. My orthodontist reassured me that it wasn't a big deal – I'd just passed out and he'd seen several patients do the same. However, when I met with my team an hour later, they all disagreed, and in hindsight, I'm thankful they did. One of them frowned as he said, "You're acting really weird – maybe you should go see a neurologist."

That didn't make sense in the moment, as I thought I was in perfect health. Two months earlier, I'd finished an Ironman and just four days before my blackout, I had climbed Haleakalā in Maui on my bike. But with my health evidently in jeopardy, I heeded my friend's advice and went to UCLA Santa Monica Medical Center's ER. After I explained my symptoms, the doctors rushed me in for a CAT scan.

They told me that I likely had a cavernoma, a benign series of blood vessels that had ruptured in my right temporal lobe. My out-of-body experience in the orthodontist's office had actually been a seizure. A nurse told me this and then said, "We can't do anything for you here and need to move you to the ICU at Ronald Reagan UCLA Medical Center." I started crying, and she held me as if comforting a lifelong friend. She was my world in that moment and reassured me that I would be ok.

Still, there was no time to waste. The morning after being rushed to the ICU at Ronald Reagan, my doctor conducted an angiogram to see exactly what was wrong. It confirmed the cavernoma diagnosis, which meant that I needed to have surgery six weeks later. On January 8, 2020, a surgeon removed my skull cap and cut a hole in my brain two inches behind my right eye. He then removed the cavernoma, put in a titanium plate that I'll have in my head for the rest of my life, and sewed me back up. During the next three days as I slipped in and out of consciousness in the ICU, I gained insight into what fate had befallen me. As a result, I'd awoken with a new perspective, a new timeline, and a new mission.

Creating a Censorship-Resistant Utility

After I was released from the ICU, I had an overriding desire to move from California to Austin, Texas. I had only two friends there, but some force I couldn't see was pulling me. So I followed that impulse and left everything I knew behind in SoCal. Once settled in Austin, I started diving deeper into online censorship and met several acquaintances who would become business partners a few months later.

Our new venture extended beyond my previous company's efforts to connect creators and brands directly without a third party in the middle. After I clarified my mission of defeating censorship, I realized that a combination of Bitcoin and the Lightning Network offered the best and only option to achieve this. The further I went down the Bitcoin rabbit hole, the more convinced I became that this potent technology would enable my team to create a decentralized, fully encrypted utility that was censorship-resistant, ad-free, and didn't retain – let alone sell or manipulate – private customer data for profit. And we wouldn't have "users"

– which are what drug pushers call their addicts – but rather members.

During this fact-finding period, I was introduced to the man who wrote the foreword you've already read: JP Sears. He went even deeper than my friend Keegan in explaining why creators' livelihoods and freedom of expression were under threat from social networks, search engines, and centralized technology and banking systems. JP was also frustrated that fans/patrons often couldn't support the artists whose work they value. Why? Because such individuals get deplatformed or canceled by the whims of the woke mob and biased employees of the tech companies that control the flow of information and reimbursement between content creators and consumers.

Reclaiming Digital Sovereignty

Once I'd clarified what I wanted to achieve with JP Sears and found early investors like Tony Robbins and Aubrey Marcus – who were just as passionate about protecting your digital sovereignty – it was a question of figuring out how to accomplish my mission. Initially, I got laughed out of more meetings than I can count by people who thought our goal was impossible. Surely Bitcoin is just useful for sending and receiving cryptocurrency payments without needing a bank, credit union, digital service like PayPal, or platform like Patreon to mediate the transaction, right?

Wrong, as it turned out. There is Bitcoin the asset and then Bitcoin the network (and in particular, the Lightning Network). The former is a censorship-resistant monetary exchange that changed everything and made cryptocurrency viable. This peer-to-peer financial network empowered us to build a creator-led utility called Zion that bridges the artist-to-fan relationship with smart contracts.

INTRODUCTION

Don't worry, this isn't about to turn into a tech textbook that only computer science PhDs could decipher. But this is a book about freedom. And specifically, freedom through encryption. In the transition from centralized, platform-governed networks to decentralized, peer-governed ones built on the Bitcoin standard and Lightning Network, you'll reclaim your digital sovereignty from the unelected and unaccountable social media barons and their underlings who serve as the arbiter between creators and audiences.

In this new age, you can be a fan who freely connects with creators and forges the same kind of authentic one-on-one relationships that you have with your friends and family, in a totally secure and private online environment in which you will never be manipulated, productized, or exploited for profit or any other reason. And vice versa if you're a creator.

Writing a New Social Media Manifesto

We believe that future social networks must:
1) Be built on a decentralized monetary layer
2) Allow for permissionless innovation, which means it's Open Source
3) Be focused on Peer Governance vs. Platform Governance
4) Be censorship-resistant at the protocol layer
5) Allow creators to own everything
6) Give every member of the network digital property rights through encryption

We've fast-forwarded to the present day at a pretty rapid pace, in both my personal how-we-got-to-now story and the new age of decentralized, censorship-resistant digital sovereignty through encryption. In the coming chapters, we'll:

- Compare and contrast the centralized century that I was born in with the decentralized era
- Examine the current state of censorship

INTRODUCTION

- Share how platform governance is transitioning to peer governance
- Explain why silencing artists is a threat to our democratic values
- Demonstrate how the new Bitcoin standard will change our world

Joining the Fight for Freedom

As you prepare to move onto the next chapter, I present Morpheus's rallying cry from *The Matrix Reloaded*:

> Zion, hear me! It is true, what many of you have heard. The machines have gathered an army and as I speak, that army is drawing nearer to our home. Believe me when I say we have a difficult time ahead of us. But if we are to be prepared for it, we must first shed our fear of it. I stand here, before you now, truthfully unafraid. Why? Because I believe something you do not? No, I stand here without fear because I remember.

I remember that I am here not because of the path that lies before me but because of the path that lies behind me. I remember that for 100 years we have fought these machines. I remember that for 100 years they have sent their armies to destroy us, and after a century of war I remember that which matters most. We are still here! Tonight, let us send a message to that army. Tonight, let us shake this cave. Tonight, let us tremble these halls of earth, steel, and stone, let us be heard from red core to black sky. Tonight, let us make them remember, THIS IS ZION AND WE ARE NOT AFRAID!

CHAPTER ONE
THE CURRENT STATE OF CENSORSHIP

> "If liberty means anything at all, it means the right to tell people what they do not want to hear."
>
> **GEORGE ORWELL**

Deplatforming. Shadow banning. Terms of service violations. Such words and phrases weren't even a thing until the past couple of years, when a handful of social media moguls decided to divide us like the Biblical analogies of sheep and goats or wheat and chaff. In other words, the content you post is either desirable, making you part of a *Mean Girls* in-crowd, or undesirable, making you persona non grata. You might have freedom of speech but no longer the freedom *after* speech consecrated in the First Amendment. Let's pull back the curtain to reveal the true motivations behind censorship, its grave consequences, and a potential path forward.

When Artists are Muzzled

In every society throughout history, artists have served as cultural consciences, truth-tellers, and the voices of people who cannot speak for themselves. Which is why tyrannical regimes have always sought to silence those who refuse to become propaganda-spouting puppets. In the wake of the Communist takeover that followed the Russian Civil War, the old power structures were replaced by a system in which equality was preached. But in reality, its leaders believed that "some animals are more equal than others" (as we learned in Orwell's *Animal Farm*) and acted decisively to silence these undesirable "others" in the creative arts.

To help destroy anti-Communist opposition, party leaders in Moscow ordered their state security forces to murder writers, painters, sculptors, and other artists en masse and to ship many more off to serve out hard labor sentences in the prison camps scattered across Russia's harshest landscapes. There were no trials or other forms of legal due process for such punishments – lists were drawn up, purges conducted, and exiles enforced.

Perhaps predictably, the best quote about what happens to a nation's collective consciousness when dictators silence artists comes from someone who experienced the consequences of censorship in the sub-Arctic wilds of Siberia. "Woe to that nation whose literature is disturbed by the intervention of power," Aleksandr Solzhenitsyn wrote. "Because that is not just a violation against 'freedom of print,' it is the closing down of the heart of the nation, a slashing to pieces of its memory." Creators should be granted true freedom of expression, Solzhenitsyn asserted, because "world literature has it in its power to help mankind in these its troubled hours, to see itself as it really is, notwithstanding the indoctrinations of prejudiced people and parties."

The "national heart" that the Russian writer referred to was also made to beat to the tune of a dictator's drum in Nazi Germany. Chief propogandist Joseph Goebbels ordered the May 1933 burning of books that were deemed to be "objectionable." Then came the 1934 law forbidding newspapers, magazines, radio stations, film studios, musicians, and book publishers from distributing content that went against the Nazi Party's "official" proclamations issued by the Propaganda Ministry. In his book *The Boys in the Boat,* Daniel James Brown quoted a University of Washington dean's observation of Hitler's realm in the mid-1930s: "The people of Germany today are afraid to express opinions even on trivial matters." Brown went on to state that "anyone saying anything that could be interpreted as unflattering to the Nazis was liable to be arrested and incarcerated without trial."

The year after the media censorship legislation came into force, Leni Riefenstahl's horrifying propaganda film *Triumph of the Will* showed Hitler bellowing to hundreds of thousands assembled below his dais at the Nuremberg Rally, "We want our people to be obedient, and you must practice obedience! Before us Germany lies. In us Germany burns. And behind us Germany follows." And as der Führer, Goebbels, and their henchman knew all too well, such obedience could only occur if dissenting creators were silenced – in some cases, permanently.

The Cheese is Only Free in a Mouse Trap

While comparisons to Nazi and Stalinist totalitarianism might seem extreme and today's outspoken artists are not – thankfully – being sent to concentration camps or the gulag, they are still combating censorship. Let's look at a few reasons why the digital state suppresses certain people, while allowing others to continue using their services unfettered.

The raison d'etre for centralized social media companies as for-profit businesses is to make money, satisfy shareholders and investors, and keep customers happy. As it's your time, focus, and energy that's being extracted and monetized, you are the product. Advertisers are the real customers because they're the ones paying social media platforms to mine your attention.

Whenever creators share content that goes against the left-leaning mainstream narrative, it's bad for LinkedIn's, Twitter's, Pinterest's, YouTube's, and Meta's business. Let's say there's a large advertiser that spends millions a year on social media ads, and they say, "You know what, Meta, we don't want to be associated with this kind of creator. You need to stop them posting these types of things on Facebook and Instagram or we will take our advertising money elsewhere." So Facebook decides that they're going to delete posts, "shadow ban" people, temporarily suspend their accounts – or remove them from the platform permanently. This could be accomplished through algorithms that automatically decide which content to elevate and which to suppress, the biased decisions of employees, or both.

Meta justifies such censorship by claiming that the users' posts have violated their terms of service and then, if the ban becomes controversial because some of these individuals have large followings, connects this verdict to some convenient political agenda item. But the true motivation underpinning social media bans is safeguarding ad revenue because, as Facebook's founder/emperor told policymakers in 2018, the only way that Meta keeps its platforms free is by raking in billions of dollars from online marketing. Which in turn is how many companies drum up a large portion of their annual revenue (see the "brought to you by Pfizer" video that went viral in fall 2021).

Cast Out into Digital Darkness

Telling the truth comes at a heavy cost these days – at least if your definition of what's true runs contrary to the stories spun by the White House, Big Pharma, and Big Tech to protect their control-profit industrial complex under the guise of "health and safety." If you shout loudly for long enough through any centralized social media platform in a manner deemed incompatible with their advertisers' interests, you might have a tweet or two deleted. Should you continue your transgressions against arbitrary terms of service, you'll probably be shadow banned. Next up is a temporary suspension of your account. Still haven't learned your lesson like you've been indoctrinated at a re-education camp? Kiss your online platform goodbye, my friend, because you're about to be deplatformed.

This doesn't just happen to everyday Joes and Janes. Can you recall the kerfuffle caused when Nicki Minaj, Aaron Rodgers, Draymond Green, Kyrie Irving, and Matthew McConaughey dared to question the Fauci-Pfizer edicts on COVID-19 vaccination? Instant vilification followed from so-called "news" anchors, the Twitter woke mob, and cultural commentators. As unpleasant as the blowback was for these individuals, it paled in comparison to Parler's wholescale annihilation. The nascent social media network was taken out at the knees in January 2021 for over a month by the world's largest online hosting platform and the two main app stores. Imagine that – a whole platform being dismantled with the casual flick of a left-leaning finger.

These were mere sideshows compared to what occurred in the big top tent of the centralized, ad-driven social media circus. Having tried and failed to get Donald Trump impeached, his opponents changed tactics and took their fight from the corridors of power to Facebook and Twitter's fiefdom. In the wake of the Capitol riots – a spark that was fanned into a raging inferno by false claims made by a certain three-initialed politician and then repeated ad infinitum in web-based echo chambers until they became public record – the media needed someone to blame. And who better to take the fall than the focus of the Left's fury ever since he declared his candidacy for the 2016 election – the 45th president.

Within weeks, Twitter had acted as police, judge, and jury as it accused, tried, and convicted Trump for violating its terms of service by "inciting violence." Facebook and Instagram soon followed suit. And so it was that the acting president of the most powerful nation on the planet lost his online voice. All because he violated a set of guidelines drawn up in Silicon Valley and enforced by Big Tech billionaires who, unlike their target, are both unelected and unaccountable.

Taming TV and Print Media

This kind of censorship isn't confined to social media but extends to traditional mediums too. In November 2021, former Fox News host Jedediah Bila appeared on *The View* to promote her new book, *Dear Hartley*. As she explained her personal decision to not get the COVID-19 vaccine, the host decided that enough was enough, and cut the video. Her justification to millions of Americans watching at home: "I just really don't think that we should allow this kind of misinformation."

Despite having much stronger credentials than his armchair critics, Dr. Robert W. Malone, who contributed to the creation of the mRNA vaccine technology and is an outspoken opponent of the CDC and FDA's COVID-19 policies, was also suppressed by social media services and banned from even *reading* articles published on *The New England Journal of Medicine* website. And what of newspapers? As George Orwell once stated, "All the papers that matter live off their advertisements, and the advertisers exercise an indirect censorship over news."

Building a Better, Censorship-Resistant Network

We've just explored how today's centralized social and traditional media services are built on a paid advertising foundation that is susceptible to free speech-defeating censorship, extreme political bias, and advertiser demands about perpetuating mainstream groupthink. Therefore, if we're to consider what a censorship-resistant network of the future could look like, we have to start with removing advertising altogether.

The only commercially viable way to run an ad-free platform is for members to pay for its use, much like they would for a subscription to Netflix or Dollar Shave Club. Their monthly or annual payment would not only free them from the behavioral nudging influence of advertising but would also win them their own node (aka a virtual server with a single user). Such nodes are inherently censorship-resistant, as they prevent the company providing the platform from censoring, banning, and deleting content that they find unpalatable.

If such a network was built on the Bitcoin standard, employees would be prevented from mining members' personal data, utilizing it to game buying behavior, or selling it to nefarious third parties because of digital encryption. In fact, the company's staff wouldn't even see this information. This means that there's no possibility of repeating the Parler debacle, the ostracization of Aaron Rodgers, Nicki Minaj, and Kyrie Irving, or the deplatforming of the president of the United States.

Such a shift to censorship-resistant networks is an example of decentralization and the evolution of digital property rights. If you're on current social media services, it's like you're living in a cabin in someone else's backyard without a rental agreement. As long as you play by the rules, you're welcome to stay. But as soon as you go against the tenancy agreement by having a conversation with your friends that doesn't align with the mainstream narrative, they can and will evict you, while keeping all your furniture, clothes, and belongings. On a new, censorship-resistant platform, you build that same cabin on land that you own, so when you invite friends over to talk, you're free to say whatever you want. And everything in your home is yours forever.

CHAPTER TWO
THE CENTRALIZED VS. DECENTRALIZED CENTURIES

"The only way to control chaos and complexity is to give up some of that control."

GYAN NAGPAL,
TALENT ECONOMICS[1]

In the previous chapter, we took a long, hard look at the current state of digital censorship in the age of COVID-19. Now let's explore one of the mechanisms that enables such repression of free speech to flourish: centralization. Before we get to centralized online and social media, we first need to trace a brief history of centralization in something that's essential in all our lives: money.

The Gold Standard

As we'll soon see, the years between 1900 and 2010 are the true centralized century and a bit. But this isn't to say that centralization wasn't a thing beforehand. The Roman Senate, feudal system, Catholic church, British Parliament, and just about any armed service are all examples of top-down power being consolidated in one central body that controls a network of subsidiary organizations below it.

There are plenty of valid reasons for centralization. It allows those in power to bring together disparate groups, provides structure to previously chaotic or disorganized ways of doing things, and enables rules to be created and enforced. Centralizing something also brings a level of standardization. Let's examine how this applies to the monetary system.

The ability to pay for goods and services is fundamental to our development as a species. At first, warring tribes would plunder the peoples they conquered and take the spoils of battle for themselves. Then came the barter system, in which I gave you my goat in exchange for a year of free eggs from your chickens. That worked fairly well but was rather arbitrary, and so came coined currency. This assigned a pre-determined value to a piece of precious metal of varying sizes, thicknesses, and weights.

Coinage came to bear the image of the ruler of the day – Caesar Augustus and other Roman emperors, Queen Elizabeth for all coins circulating around the British Empire, and so on. And with money came taxation, largely because governments needed to pay for colonialization, wars, and development of public amenities like roads, bridges, and aqueducts without relying on the personal fortunes of kings, queens, and emperors.

In theory, a government could just print as many coins as it wanted. But market forces come into play, and currency can become devalued if there's too much of it. Also, ruling authorities soon realized that when it came to large transactions such as loans or war reparations, they couldn't just hand out IOUs. There needed to be enough money or assets in their treasury to actually pay.

Enter the gold standard, ushered in by the German government in the wake of the Franco-Prussian War (aka the War of 1870), adopted by other countries in 1871, and widespread internationally by 1900. This tied the production and distribution of currency to one of the sturdiest sources and symbols of wealth throughout the millennia – gold. If there wasn't enough of it in a government's coffers, it could no longer issue currency because, as American financial tycoon JP Morgan later remarked, "Gold is money. Everything else is credit." There were no trillion-dollar infrastructure bills being passed during the Victorian era because there wouldn't have been sufficient gold reserves to justify the expenditure. The gold standard was only abandoned at the start of the Great Depression, when governments decided to artificially stimulate the economy.

Centralized Banking After the Wall Street Crash

It's hard to overstate the ramifications of the Great Depression, which was, by every measure, infinitely worse than the Great Recession we experienced between 2007 and 2009. By the mid-1930s, 50 percent of mortgages were in default, unemployment had soared to 25.6 percent, and millions were left to live in so-called "Hooverville" shanty towns. To make matters worse, the bitter harvest of over-farming and drought – the Dust Bowl – devastated American agriculture. During his four-year tenure as New York governor, Franklin Delano Roosevelt (FDR) put programs into place to try and alleviate the economic misery. These served as a proving ground for the schemes that he would extend nationwide after defeating Herbert Hoover in the 1932 presidential election.

From the moment he took office, FDR followed the example of Great Britain, which had abandoned the gold standard in 1931, and took America down the path of a new centralized banking era. In 1933, the president's Executive Order 6102 demanded that all holdings in local reserves must be transferred to the US Treasury – a massive consolidation and centralization of wealth. The 1934 Gold Reserve Act that he signed into law the following January gave him the power to devalue gold, and with the next stroke of his pen he did just that, wiping 40 percent of its value away.

FDR then increased interest rates to elevate the demand for dollars and ushered in the age of "tax-and-spend liberalism." Upper, middle, and lower classes were all required to fund new federal programs like Social Security with mandatory contributions from each paycheck. The top marginal income tax rate skyrocketed from 25% to 63% in 1932 and to 79% in 1936, while the lowest earning Americans were taxed over 10 times more in 1932 than in 1929 (four percent versus 0.375 percent).

These huge tax hikes coincided with the US government tightening the reigns on the supply of dollars into the domestic and global economy. As Nik Bhatia writes in *Layered Money*, "the Federal Reserve finally secured its official monopoly over note issuance… Thus, began the journey for the US dollar to stand alone, independent of gold."[2]

Nine years later, senior US Treasury official Harry Dexter White led negotiations with British economist John Maynard Keynes and officials from other governments to determine the future of global currency exchange. "By the end of the conference, White had outmaneuvered Keynes to establish a global financial framework with the US dollar firmly at its core," a summary of Benn Steil's *The Battle of Bretton Woods* states.[3]

The Rise of FDR's Three-Lettered Agencies

It would be false to state that FDR was the first US president to establish a centralized federal agency. His cousin Theodore (Teddy) Roosevelt created the United States Forest Service (USFS) and established 150 national forests, 51 federal bird reserves, 18 national monuments, five national parks, and four national game preserves through the 1906 American Antiquities Act.

That same year, Teddy signed the 1906 Food and Drugs Act into law to regulate so-called "patent medicines" (the forerunner of what we now call supplements, many of which were useless or harmful "snake oil") and stop harmful food production practices. This was a time when rotten meat was being packaged at slaughterhouses and damaging ingredients like lead paint and heavy metals were widely used as food colorings. Roosevelt's Food and Drugs Act tried to put a stop to such practices through the new Food and Drug Administration (FDA). The former Rough Rider colonel also led the charge to limit the power of the "robber barons" like the Vanderbilt family, utilizing the 1902 Sherman Act to break up the monopolies that had a stranglehold on many American industries. Before he lost the Republican nomination to his protégé William Howard Taft, the Federal Bureau of Investigation (FBI) was created during Teddy's presidency.

As monumental as such legislation was, it paled in comparison to the Democratic Party's prolific lawmaking three decades later. FDR's presidency ushered in so many centralized government entities that they soon were grouped with the term "alphabet agencies." Wikipedia's partial list contains 37 of them.[4] Some, like the Agricultural Adjustment Administration (AAA), Farm Credit Administration (FCA), and Drought Relief Service (DRS) sought to alleviate the suffering caused by the Dust Bowl, while many more – including the Civilian Conservation Corps (CCC), Civil Works Administration (CWA), and Works Progress Administration (WPA) – tried to put America back to work in a time where jobs were scarce and the economy had been brought to its knees by the Great Depression and the Dust Bowl.

Though the intentions of FDR and his colleagues on Capitol Hill were likely noble, it was still the most intensive period of Big Government expansion in US history, and in just five years (1933 to 1938), FDR created four times more federal agencies than in all the years since. That includes the period after the 9/11 terrorist attacks when George W. Bush instituted the Transport Security Administration (TSA), Immigration and Customs Enforcement (ICE), and the Department of Homeland Security (DHS). FDR's spate of three-letter agencies marked the apex of the centralized century, with Washington DC officials overseeing financial services, air travel, employment, food production and distribution, home ownership, utilities, shipping, conservation, and many more aspects of American life.

iPhone and Facebook
Usher in the Decentralized Century

While the new century technically began on January 1, 2000, it's arguable that the decentralized century actually started with the US release of the first Apple iPhone on June 29, 2007. Not only did this revolutionary device put the power of a personal computer in your pocket, but it also enabled you to create your own application and sell it to millions of users around the world via the App Store in a way that was previously only accessible to large technology companies through traditional retail channels. As a purchaser of such apps, you could join dedicated communities of fellow users who were into fitness, hiking, birdwatching, and so on. This was another decentralized step in the right direction.

Soon enough, Samsung, Motorola, and other device manufacturers had begun selling iPhone imitators, many of which ran on Google's Android platform. This too offered its own app marketplace – now called Google Play – which enabled you to publish your own app or use other people's. Another way that cell phones running on both major platforms also decentralized communication was by allowing you to text friends with individual or group messages. Soon enough, dedicated apps sought to enhance this, such as Slack for workplace communication and Voxer for audio-based conversations.

These products and services were hallmarks of the decentralized century that gathered pace in recent years. But just as it looked like you might finally be free to communicate with your buddies, family members, and colleagues on your mobile devices, Big Tech started trying to claw back some of the ground it was losing or couldn't further commercialize. You could be under the impression that what you said on phone calls, wrote in texts, and shared via app-based messages was really private, but there are provisions in the Patriot Act for government entities to monitor, track, and record such communications. In addition, social media companies and app developers claim to adhere to the latest security and privacy standards, but in reality, if you're using their stuff, your data is fair game to be analyzed, logged, and resold to third parties (Don't believe me? Check the fine print of these apps' terms of service).

At the time of writing this chapter, there were even suggestions that Big Tech companies would soon start censoring texts and other messages that "spread misinformation" about COVID-19 treatment. We have also heard a true, firsthand story of a friend's app being denied entry to the App and Google Play stores because he offered breathing techniques that could help people recovering from the coronavirus. Somehow this is deemed to be "misinformation" because helping people with non-vaccine protocols cannot benefit the Big Pharma companies who spend so much money on marketing through both traditional and social media platforms.

In these ways, we are living through a backlash to decentralization, in which Big Tech and the marketers who fund their increasing expansion are trying to revert back to all the old habits of centralization. These include censorship, bureaucracy, overreach, and interference in communication and commerce between consenting parties. Centralization offers those holding the puppet strings the kind of control, monetization opportunities, and chance to force their own personal and political bias upon the consumers who they so callously call users.

When the decentralization that began with the introduction of the iPhone threatens their hegemony and profit margins, these puppet masters start taking decisive actions to undermine whatever progress has been made and reassert their dominance and authority. This is why the journey from the centralized era into the decentralized age isn't linear but is more of a back-and-forth process fraught with detours, potholes, and diversions.

Bitcoin Cracks the Centralized Transaction Model

In the first half of this chapter, we looked at how in the 1800s, 1900s, and early 2000s, money supply, circulation, and regulation was centralized – first through the gold standard and then the consolidation of central banking that came after it was abandoned. Now let's see how cryptocurrency is starting to turn the tide toward decentralization. The magnitude of the financial crisis that unfolded between 2007 and 2009 suggested that the global monetary system wasn't securely anchored by a firm foundation as investors had long been promised. It really sat on pillars made of sand that could be blown away as soon as subprime mortgages started defaulting at a certain rate (see Michael Lewis's book *The Big Short*).

The aftershock resulting from the fiscal earthquake shook every country's financial system to its core. Perhaps this is why when he mined the first Bitcoin block on January 3, 2009, the individual who goes by the alias Satoshi Nakamoto embedded a line of code in the service's permanent ledger that read, "The Times 03/Jan/2009 Chancellor on brink of second bailout for banks." He was referring to a story in the UK's broadsheet newspaper detailing that the chancellor of the Exchequer, who oversees the British government's budget and economic policy, was preparing to prop up the huge multinational banking corporations whose malpractice and greed had imperiled the livelihood and financial security of the entire world's population.

A few weeks later, Nakamoto went on to explain other reasons why his loss of faith in the centralized banking system had led to his creation of Bitcoin, which he defined as "A purely peer-to-peer version of electronic cash [that] would allow online payments to be sent directly from one party to another without going through a financial institution."[5] Nakamoto wrote on Bitcointalk that "The root problem with conventional currency is all the trust that's required to make it work. The central bank must be trusted not to debase the currency, but the history of fiat currencies [those issued by decree – the Latin meaning of fiat – by government-run central banks] is full of breaches of that trust. Banks must be trusted to hold our money and transfer it electronically, but they lend it out in waves of credit bubbles with barely a fraction in reserve."[6]

What Nakamoto hinted at here was that throughout the history of monetary networks, banks have let their trusting customers down. The most dramatic example wasn't the Great Recession, as most deposits had long since been guaranteed by FDIC insurance, whereby the US government promised that money paid into financial companies (or, at least, those who subscribed to the FDIC standard) would not be lost even if that bank failed or its stock price decreased. No, the biggest banking betrayal in US history actually occurred during the Great Depression when banks did not have the assets on hand to pay out customers who wanted to withdraw their money. So the everyday folks who had trusted their local bank with their life savings were left empty-handed.

Nakamoto saw that while this didn't happen anymore in literal terms, the misdeeds of big banks like Deutsche, Bank of America, and Morgan Stanley had again plunged their customers into the mire of a global recession, leading to widespread unemployment, the loss of retirement investments, and much more misery. He envisaged Bitcoin as a way to cut out the middleman – the bank – and enable users to complete peer-to-peer transactions using a more stable digital asset.

He also stated that it could potentially provide "the 'high-powered money' that serves as a reserve currency for banks that issue their own digital cash."[7] In short, Bitcoin could become both a new, better kind of gold standard and also a decentralized layer for money storage and transfer that wasn't subject to the bad banking practices that had screwed customers out of their savings and investments during the Great Depression, Great Recession, and other financial downturns. With Bitcoin, there was no longer a need for a human mediator between two individuals who wished to trade – they could complete such transactions securely, anonymously, digitally, and with smart contracts that protected both parties.

Social Media Decentralization

On September 4, 2021, angel investor and entrepreneur Balaji Srinivasan, formerly the CTO of Coinbase and general partner at Andreessen Horowitz, tweeted, "The Decentralization gave rise to the ongoing Counter-Decentralization. But the Counter-Decentralization will give new energy to the Decentralization. Why? Because millions now realize that all those theoretical risks of centralization are very real."[8] This was part in response to his own tweet posted in February 2018, in which he stated that "Censorship incentivizes decentralization."[9]

Online censorship was certainly one of my main incentives for creating a decentralized social media utility, as was enabling freedom of and after speech, decoupling user content from an advertising-centric model that's subject to the whims of big marketing budgets and employee bias, and empowering creators and fans to connect and compensate each other directly.

To us, the risks of centralization and the over-concentration of power, influence, and reach in the hands of a few technology robber barons isn't merely theoretical: its negative impact is being felt far and wide as billions of users are manipulated, censored, and prevented from expressing unfettered opinions that differ from the macro-narrative spun by the propogandists employed by Big Tech, Big Government, and Big Pharma. That's why we sought to reclaim such power and put it back in the hands of the very people who should've wielded it in the first place: you and the creators you want to connect with and support.

Decentralized networks don't merely make it hard for censorship to exist – it is literally impossible because they don't even see your data, let alone store, commoditize, or report on it. By utilizing a subscriber-based pricing model, such companies are supported by memberships, not by propping up the charade of free use with advertising revenue that extracts your time, attention, and energy on other centralized platforms.

THE CENTRALIZED VS. DECENTRALIZED CENTURIES

Another way that decentralized networks give you a better experience than Meta, Twitter, LinkedIn, TikTok, and the rest owning your content by default – simply put, if you use these platforms, you're agreeing to this – they don't lay any claim to your content. Whatever text, videos, photos, etc. you post are yours. In this way, you're given back true and total ownership of everything you share today, tomorrow, and next year. And unlike how Bane and Miranda Tate continue to rule Gotham with a fascist, lockdown-controlled rod of iron after he declares, "Gotham is yours – do as you please," in *The Dark Knight Rises*, you actually are free to do and say what you like on decentralized utilities. In the next chapter, we'll explain more of how this decentralization and the creation of a censorship-resistant platform benefits you and your community.

CHAPTER THREE
PLATFORM VS. PEER GOVERNANCE

"It is hard to imagine a more stupid or more dangerous way of making decisions than by putting those decisions in the hands of people who pay no price for being wrong."

THOMAS SOWELL

In the previous chapter, we explored the ongoing transition from the centralized banking and technology that dominated society until the introduction of the iPhone and Bitcoin, the decentralized ones that these two innovations have ushered in, and how those powerful people in charge of the former are trying to stifle the progress of the latter.

Now it's time to turn our attention to another piece of this digital puzzle. In this chapter, we'll examine why control-oriented, free speech-averse platform governance is a hallmark of centralized networks, while greater civility, accountability, and censorship-resistance can be found through a peer-governed model backed by digital credit.

What is Platform Governance?

Platforms are a group of monetized technologies that act as a foundation of development for other applications, while governance can be defined as the rule by some over many. At its essence, the term platform governance is a catch-all for the rules, guidelines, and regulations that influence how content is uploaded, stored, and disseminated via a particular product ecosystem. The most popular centralized social media networks have variations on this theme, but make no mistake about it, they are all prime examples of platform governance in action.

Once you create an account on such a centralized network, you're joining this company's ecosystem, and they have oversight of everything you do from that moment forward. As Mark Zuckerberg himself reluctantly admitted when he was questioned by a Senate subcommittee, this extends to browsing outside of Facebook.[10] This explains why if you or a family member have been recently researching the best new running shoes on Google, you'll see ads for them from online retailers next time you log into Facebook.

Keeping you within a company's ecosystem is often characterized through the simple analogy of playing in a sandbox. But unlike a carefree child on a beach, the sand, pail, shovel, and ocean are the property of the company. And if they believe that you've misbehaved, they'll play the role of overbearing parent, order you to leave the sandbox, and never allow you back on that beach.

Everything that you do within the company's ecosystem from day one is subject to the control of their community guidelines. Similarly, each piece of content you upload, including photos, videos, and text-only posts, is owned by them. As is every single kilobyte of your data and even your behavioral patterns when you're using the system (like how long you look at other people's posts and the likes, comments, retweets, and other ways you interact with them), the combination of which drives both advertising and the way that certain content is elevated or relegated in your feed.

When you participate in a centralized, platform-governed ecosystem, you automatically surrender your digital property rights. Yes, the platform is technically free to use, but you're subsidizing this by allowing the owners of the company to extract your time, attention, and – as is crucial to the success of their ad revenue-based business model – money. These are the deviously hidden costs of surface-level entertainment and connecting with other users.

By its very nature, platform governance is built on terms of service. You might have never been aware of these as you go about your online business, but the very use of those applications means that you agree to abide by this rulebook, the same as a professional basketball player indicates his consent to the NBA's governance every time he steps on the court. If he fouls an opposing player too hard, he can be assessed a flagrant one penalty. If it's a deliberate shot to the head or groin, then he will likely be given a flagrant two, resulting in his ejection from the game.

Should he or his coach dare to question the officiating after the game, they will be fined. It doesn't matter if the individual thinks their punishment is unjust because the NBA rulebook is sacrosanct and all-powerful. The key difference here, though, is that pro basketball players are paid substantial sums to compete, while you're just trying to interact with people and connect with creators whose work you admire.

Platform Governance Turns Tyrannical

If the ruling authority of a centralized platform – in this case, the company and its employees – deems you to be in violation of any of their arbitrary rules, they can take whatever actions they like to punish or censor you, mostly because you have no digital property rights. The term "my (insert social network)" is factually incorrect because you have no ownership. We've all seen the ill effects of such repercussions, as freedom of and after speech was suppressed during the height of what author and former *New York Times* reporter Alex Berenson calls the COVID-19 "pandemia." As you've read in a previous chapter, this also applied in the political sphere when Donald Trump was unceremoniously ejected from all major social media networks. Such are the risks of expressing your own opinion in platform-governed ecosystems.

The removal of posts and people's accounts are just two of the most overt punishments passed down from the rulers of the platform-governed digital kingdoms. Your UX (user experience) can become slower, while certain posts are buried to make it harder for your followers to find and share them. At its core, platform governance is all about the control of the many by the few, the limiting of expression that goes outside the lines of what the social media magnates deem to be advantageous to their causes, and the preservation of the status quo. Dare to contradict any of these inflexible tenets, and you should get ready to be muzzled, censored, and muted.

Any time they are confronted by accusations of bias – whether by lawmakers, journalists, or proponents of humane technology like Tristan Harris and Jaron Lanier – social media company executives fall back on the claim that their services are neutral. In his booked *Zucked*, former adviser to Mark Zuckerberg, Roger McNamee, took issue with this and wrote, "While Facebook argues that its technology is 'value neutral,' the evidence suggests the opposite. Technology tends to reflect the values of the people who create it. Algorithms are used throughout the economy to automate decision making. They are authoritative, but that does not mean they are fair."[11]

In addition to its actual lack of fairness, another pitfall of platform governance is that the companies who rule their ecosystems have made themselves the de facto arbiter of truth. You've witnessed this in action if you've ever seen the term "fact checked" on a post or Google search results. Despite their pretense that the investigation of a supposedly suspect piece of content is being conducted in the name of truth and accuracy, the inconvenient truth is that Meta, Google, and other all-powerful tech companies are paying the "content moderation" companies who issue such rulings. And not a pithy sum either, but rather hundreds of millions per year – an amount that is only increasing as they seek to consolidate their control of how politics, COVID-19, and other hot topics are referred to. Why? To further their own political aims and satisfy the demands of their advertisers.

Contrary to what your platform-governed network overlords would like you to believe, the censorship resulting from the "fact checking" process has very little to do with whether something is true or false and is in no way "independent." They are simply opinions. As a case in point, Fiona Godlee and Kamran Abbasi, editors of the renowned *BMJ* (formerly known by its longform title, the *British Medical Journal*), wrote an open letter to Mark Zuckerberg in December 2021 taking issue with the way in which one of the companies Meta paid had censored a *BMJ* article published the previous month:

> From November 10, readers began reporting a variety of problems when trying to share

our article. Some reported being unable to share it. Many others reported having their posts flagged with a warning about "Missing context ... Independent fact-checkers say this information could mislead people." Those trying to post the article were informed by Facebook that people who repeatedly share "false information" might have their posts moved lower in Facebook's News Feed.

Group administrators where the article was shared received messages from Facebook informing them that such posts were "partly false." Readers were directed to a "fact check" performed by a Facebook contractor named Lead Stories. We find the "fact check" performed by Lead Stories to be inaccurate, incompetent and irresponsible.[12]

Godlee and Abbasi have every right to be outraged by the way Meta censored the *BMJ* piece in question. As they stated elsewhere in their letter to Zuckerberg, the article had already been evaluated before publication by credentialed research scientists who were experts in the author's field. As this peer review and the simultaneous pre-publication legal review were thorough and met the high standards of a reputable journal, there was no need to have a fact checker with more bias than domain expertise assess it further.

This fact checking only took place because the content and conclusions of the paper in question ran contrary to the public statements of Dr. Anthony Fauci, the CDC, and the Biden administration. Therefore, it was potentially undesirable to Meta advertisers who wanted their content to run alongside posts that adhered to or perpetuated the party line. Hence the "fact check" label, suppression of the story, and erroneous verdict that the article was nonfactual. We can only hope that further challenges to the hegemony of the money-motivated, politically-biased, platform-governed networks will be made as more people push back against the unjust censorship of the material they share in good faith.

Supercomputers vs. Human Brains

A further problem with platform-governed networks is that they've created a David and Goliath-like struggle between supercomputers and your mind. As the Center for Human Technology states, social media is never neutral, and "Today's technology has increasing asymmetric power over the humans that use it. Machine learning, microtargeting, recommendation engines, deep fakes are all examples of technologies that dramatically increase the opportunity for creating harm, especially at scale."[13]

Your ancient brain has not evolved to combat the constant onslaught on your nervous system that you expose yourself to with even moderate use of platform-governed ecosystems. In an article for TrainHeroic, Kenny Kane commented on the mismatch between the almost infinite power of such technology and its impact on your mind and behavior:

> By design, Facebook/Instagram, Snapchat, Twitter, and YouTube reward the emotional fight, flight, or freeze parts of our brain, while simultaneously preventing other regions like the prefrontal cortex (PFC) from making executive decisions, like turning off the platform to do something else with your time and attention in the first place. While staring

at our screens, the experience feels important, urgent, and even good (due to the rush of dopamine and other neurotransmitters). As Adam Alter writes in *Irresistible*, the technology is designed to tap into our basic emotions and survival mechanisms and keep us there in behavioral loops rather than encouraging us to reflect and choose behaviors with greater intent.[14]

This is why you feel compelled to check your social feeds so often – there's an artificial sense of necessity creating compulsive, addictive habits. As a result, you further stimulate your sympathetic nervous system and condition yourself to be less thoughtful and more impulsive. Then you become perpetually anxious and worry that if you're not refreshing every few minutes, you could be missing out (FOMO in action). If you condition yourself to be distracted, skittish, and scattered for several hours a day and perpetuate this pattern for years on end, then is it any surprise that our society is more impatient, intolerant, and reactive than ever before? This isn't your fault but is actually the doing of platform-governed networks who are exploiting us all.

Such behavioral changes compel you to post ever more divisive and partisan comments, to gang up on people with different opinions, and to become more entrenched in your beliefs. The reason that dislike buttons and commenting features exist is not to further our powers of self-expression, but rather to denigrate them and create more division, outrage, and dissension. The behavioral psychologists employed by tech companies know full well that the more rancor there is, the more frequently you'll come back to fire another volley at your online opponents, and the longer you will spend doing it. Which all means more eyeballs on marketers' ads and greater quarterly earnings, power, and influence for Big Tech.

Beyond your primitive brain fighting a losing battle against increasingly powerful modern technology, there is a fundamental disconnect between your motivations and those of the platform-governed applications that are using you. In his book *Stand Out of Our Light*, Oxford University ethicist James Williams suggests that you probably have life-enriching aims "like 'learn how to play piano,' 'spend more time with family,' 'plan that trip I've been meaning to take,' and so on. These are real goals, human goals."

In contrast, those who control centralized social networks have nefarious targets, such as "Maximizing the amount of time you spend with their product, keeping you clicking or tapping or scrolling as much as possible, or showing you as many pages or ads as they can…these 'engagement' goals are petty, subhuman goals. No one wakes up in the morning and asks, 'How much time can I possibly spend using social media today?'"[15]

So what's the antidote to this dissonance between your goals and those of the tech bigwigs who seek to strip mine your very consciousness? In his book *The Attention Merchants*, Tim Wu issues this rallying cry. "We must act, individually and collectively, to make our attention our own again, and so reclaim ownership of the very experience of living."[16] This can only happen if you make the hard but necessary choice to walk away from Twitter, Facebook, Instagram, and the rest.

It's only by transitioning from platform-governed networks to peer-regulated ones that you can stop looking over your shoulder every time you fire up your social feeds. As a result of fact checking and other forms of censorship, we're all living in a state of fear. Balaji Srinivasan traced the history of such trepidation through three distinct stages. In ancient civilizations, people tried to avoid the punishment of their vengeful gods, whom they believed would strike them down for sinning. Then as what Dwight Eisenhower called "the military-industrial complex" reached its apex in the middle of the centralized century, the thing people were most afraid of was the nation state.

Today, we fear nothing so much as the wrath of the platformed-governed networks.[17] They have the power to ruin our reputations, take away our livelihoods, and silence our voices. In our hyper-connected world, if you get kicked off Facebook or Twitter, you've effectively been removed from public discourse. And who gets to decide this? Not you or your peers but the paid, non-independent fact checkers and their paymasters, the Big Tech barons, who are in the pockets of the ones padding their stock prices – advertisers. He who wishes to drink the king's wine must sing the king's song or will be cast out of the kingdom.

The Case for Peer Governance

In contrast to platform-governed, centralized networks, peer-governed ones involve community members abiding by a set of guidelines that they all agree on in a kind of digital charter or constitution – not those that were forcibly handed down from on high by billionaires governing by executive fiat like a power-drunk Roman emperor. On a Bitcoin-based utility, these are terms created and ratified by community participants instead of rigid terms of service dictated by a company's legal team.

While platform governance is top-down and authoritarian, peer governance is bottom-up and democratic. According to an insightful article on Resilience.org, peer governance "points to an ongoing process of dialogue, coordination, and self-organization. By recognizing individuals as active peers in a collective process rather than positioning them as adversaries competing to control a large, remote third party [or] government, a more trusted type of governance can emerge."[18]

In other words, if you participate in a peer-governed community, you have an active role to play in determining what proper usage looks like and doesn't. We'll dive deeper into how such a peer-governed, decentralized network can create a more civil and accountable public discourse in the next chapter, but for now, it's safe to say that in a platform-governed system, you and I are mere plebians or serfs, upon whom the whims, political bias, and fiscal motivations of our feudal lords can be thrust at any time. Whereas in a peer-governed model, we're all rulers of our own little kingdoms and have an important role to play in how they are run.

Srinivasan believes that making the move to decentralized, peer-governed platforms will usher in not just technological innovation but also a new age of society at large:

> Just as [British philosopher John] Locke said that the legitimate state is that state that protects property rights, encryption forms the basis of a new system of rights that undergirds the Network State. Encrypted communications networks allow not just for property rights on digital resource and information networks but also empower

rights of speech and association: physical gathering and coordination via private communication channels, tamper-proof security systems, and provable identity.[19]

Decentralized social media embodies these sentiments, as they seek to turn his vision of the Network State into "a social network with an agreed-upon leader, an integrated cryptocurrency, a definite purpose, a sense of national consciousness, and a plan to crowdfund territory" into reality.[20] It wasn't enough to take a small step for peer governance – what's needed is a quantum leap that restores your right to gather with whomever you choose and say whatever you wish without a third party silencing you or violating your right to freedom of expression. Doing so requires us to create not only a secure, ad-free, censorship-resistant network but also one in which you own your own node (aka a dedicated digital server). This ensures that you achieve full sovereignty and maintain sole digital property rights for all your content and data.

Removing the Messenger in the Middle

The way that community interaction works on a peer-governed utility is also very different from what you're used to with the current platform governance model in most Big Tech ecosystems. You can start your own community, much like you would a group on Facebook, Discord, or LinkedIn. But there's a crucial difference. To make a comment in a Bitcoin-based community, someone has to stake a certain amount of satoshis (aka SATs – a small amount of Bitcoin). If you, as the administrator, deem their input to be offensive or detrimental to the wider group, they will lose that stake, whereas if they remain civil for the duration of the staking contract period, they keep it. This way, each user is subject to the good governance of their peers, rather than to the biased, ad revenue-driven motivations of the company that owns the platform.

As each member pays a small monthly subscription fee for their own distinct node, there is no need for marketers and their agendas to dictate what is elevated or suppressed in your content feed. This isn't merely unlikely – it can't be done. Nobody else sees, stores, or attempts to monetize your data, so you can enjoy a manipulation- and censorship-proof experience that more closely mimics in-person interaction than anything you've experienced on centralized, platform-governed social media services.

Such peer governance principles are brought to life by the Bitcoin standard and the Lightning Network. We'll explain exactly how in more detail during a forthcoming chapter, but the synopsis is that we've found a way to run peer-to-peer communication up and down the same track as the one that Bitcoin transactions use. With the Lightning Network, this is instantaneous, and messages and comments cannot be deleted, interrupted, or halted because they're encrypted from end to end with a level of security that isn't possible in WhatsApp, Slack, or any other platform-governed application. There's also no central company that's routing messaging (or delaying, distorting, or burying it, as is the case with centralized social networks). It's all done through a protocol inside of a decentralized system.

We believe that digital platform governance has had its day and that just as the gold standard morphed into centralized banking and then into cryptocurrency, the way of the future is decentralized, peer-governed networks. These will allow you to reclaim ownership of your content and data, communicate with friends, family, and fans in a civil and accountable way, send and receive secure and immediate payments via the Bitcoin standard, and interact honestly without worrying about your experience being diminished by censorship, behavioral manipulation, or other actions by a company using the platform governance model.

CHAPTER FOUR
CREATING A MORE CIVIL, SECURE, AND ACCOUNTABLE ONLINE DISCOURSE

"Insidious elements have also taken hold online – the crippling insults and mockery of people, the widespread use of snark weakly justified as innocent humor, and the destruction of business reputations and individuals through shockingly easy and anonymous means."

ANDREA WECKERLE,
CIVILITY IN THE DIGITAL AGE[21]

What's the thing you like least about using social media? For me, it isn't the continual interruption of ads or the knowledge that I'm being manipulated for profit, but rather the rudeness, hatefulness, and division that has come to define online conversation. Jerry Seinfeld said that he stopped performing on college campuses years ago because some group or other would invariably protest an innocent comment he'd made in jest about something. Now such oversensitivity and outrage has become firmly entrenched in the digital sphere, and you can hardly say anything without being corrected, shamed, or slammed. Simply put, our online discourse has denigrated into a cesspool. In this chapter, we'll examine how this happened and propose a couple of ways that we can re-elevate civility and accountability to make the time you spend online more pleasant, polite, and safe.

Let me start by making it clear that I'm not suggesting there can or ever should be an absence of dissension online. As I've written several times in this book, you have an inalienable right to freedom of and after speech, and there's always going to be someone who disagrees with your beliefs, opinions, and statements. You will also encounter people in cyberspace that you don't want to become best friends with. But this being said, I hope you'll agree that we can all do better in how we interact with our peers, friends, and families online and should aim for a higher standard of common courtesy within our digital communities. I firmly believe that Bitcoin and the Lightning Network create digital trust and stability, and can help you have much richer, safer, and more dignified online interactions on new networks where it's not OK to drag others down into the mud. Those who act destructively will be held accountable.

The Causes and Consequences of Digital Discourtesy

One of the best ways to make online discourse more civil and accountable is to add a layer of accreditation inside of a digital experience. This provides a greater degree of trust for each individual within the communities that they choose to participate. In the real world, you have credit associated with your previous and current actions, whether that's an objective number like your credit score or a more subjective measurement like your reputation in your community and profession. But in the digital world, it's bifurcated because your behavior is largely separated from what you say and do. Acting out in the real world can be very high-consequence, whereas the centralized social media platforms have created a largely consequence-free alternate reality.

As we've already explored, you can be removed from Facebook, Twitter, and the other major social media services if you say anything that threatens their political or corporate agendas or is unpalatable to their advertisers. However, as the award-winning documentary *The Social Dilemma* showed, bullying, doxing, and hazing are commonplace. When combined with the comparison trap, this unsavory online behavior contributes to self-harm, dysmorphia, eating disorders, and a massive increase in teen and pre-teen suicide. An AMA article summarized these sobering statistics, which demonstrate a correlation between social media use and diminished mental health among young people: "U.S. hospital admissions for self-harm in teenage girls skyrocketed after 2010, up 62% in girls ages 15-19, and up a staggering 189% in girls ages 10-14. Suicide rates have also skyrocketed since 2010, up 70% in 15-19-year-olds, and 151% in 10-14-year-old girls."[22] Clearly, a combination of social media overuse and bad behavior online has real-world ramifications, but often not for the perpetrators.

When an individual is attacked in person, there are consequences. It could lead to a physical confrontation or the perpetrator being arrested. If the incident takes place in a sporting venue, bar, or restaurant, they could be ejected from the premises and banned from coming back. Should the police get involved, the attacker could end up with a permanent criminal record that impacts their ability to get a job.

In other words, bad behavior in the real world has tangible and, often, lasting consequences. But the digital realm remains like the Wild West, where virtually anything goes with little recourse for the victim of bullying, insults, body shaming, and other kinds of abuse. Another reason for the lack of ethical behavior on traditional social platforms is that they only require a name and email address. If you behave badly for long enough and get booted off, all you need to do to create a whole new digital identity separated from your previous actions and persona is enter a different name and email address. Then you can continue assaulting other users as if nothing ever happened.

Pro athletes, celebrities, and regular folks alike are dodging accountability and repercussions for being bad actors online with burner accounts, which they use to fire back at critics anonymously. And on platforms where entire mobs mobilize their outrage against a person whose views they disagree with, if one of the assailants gets their pitchfork taken away, the others will continue with their witch trial until the unfortunate victim is burned.

The Significance of Staking

To give you a better, safer, and more enjoyable online experience, we need to close the gap that currently exists between in-person and digital conversation. To do so, we need to build on the idea of credit inside of a social experience, which then adds a layer of censorship resistance. A decentralized utility that is built on the Bitcoin network provides safety and civility in an online world where danger, fear, and incivility run rampant on centralized platforms.

How? By asking every member to put some skin in the game and make themselves more accountable to their peers than they would be on Twitter, Instagram, and other networks. If a creator has staking turned on for the community that they administrate, a certain amount of satoshis, which is outlined in the stake contract when you join the community, is held in escrow when you make a comment.

Staking was first introduced by Paul Itoi and his team via the Sphinx app to get rid of spam. This was an evolution of the hashcash concept invented by Adam Back, which created stamps to attach to email that added a micro-cost to sending spam messages and inspired the proof-of-work algorithm that Bitcoin is based on. Itoi explained how he evolved the staking function: "Adding the ability to require a micropayment with each message prevents spam. The payment is too small for the sender to notice, but it's enough to make spam unprofitable at scale. These tiny payments are the immune cells that push out the junk content that plagues Reddit, Discord, and Twitter."[23]

Here's one way that staking can prove to be an effective deterrent against uncivil posting as well as spam. If the creator of a community in the decentralized network you're using deems your comment to be offensive or detrimental to themselves, another member, or the group at large, they'll delete the comment from the thread and you will lose your stake, as you've willfully gone against the smart contract and failed to uphold your end of the bargain by spamming, slamming, or disrespecting other members. Whereas if your comment stays on the community page for the duration of the staking contract (let's say 24, 48, or 72 hours), your satoshis will be returned without penalty.

Such accreditation and staking prevent bullying, hazing, spam, and all the other things that degrade your online experience and that of the people you interact with. It also replaces the old platform governance model with more judicious peer governance, just like you have in your neighborhood and the in-person groups that you're a member of. To regain civility in our digital discourse, there must first be more accountability. And this begins with establishing credit and consequences for actions that are harmful to individuals and communities. Staking is the simplest and most effective first step in doing so and a powerful example of how a network built on the Bitcoin standard can utilize a layer of monetary credit to encourage respectful, constructive, and considerate dialog that is often absent on platforms where accountability is lacking.

These measures can help usher in a more civilized, courteous, cross-border society in our increasingly digital world. It's how we can start to make the internet better for ourselves and our children. If your kids currently use TikTok, Twitter, or any of Meta's properties, are you ever concerned for their welfare? Check out any news website, and you'll be sure to see a story about cyberbullying and the mental and physical health issues that it causes for young people in particular.

In contrast, if they were using a decentralized utility, you could be safe in the knowledge that the comments of nefarious people would be removed and those who posted them held responsible through the use of staking contracts. No longer could your kids be subjected to verbal abuse, threats, or shaming as they can be daily on traditional social media. They'd finally have the freedom to chat with friends, connect with creators, and thrive in communities without the looming menace of having their mental health and confidence destroyed by bad actors. This is the promise that platforms built on the Bitcoin standard offer.

A Better Business Model

The use of accountability measures is just one way that Bitcoin-based tools can help clean up cyberspace and get us back to the sense of wonder we all had when first firing up the internet before those motivated by profits and one-sided politics sunk their claws into it and started stoking dissension, partisanship, and division. An interviewer once asked Sherry Turkle, an MIT professor and author of seminal books like *Reclaiming Conversation* and *Alone Together*, "In your years of studying technology, has there been a turning point, feature or service that has accelerated its unwelcome side-effects?"

"The point when Facebook and social media in general discovered its business model," she replied.[24] As we've already seen, this is selling ads. To do so for maximum profit, the platform needed your attention for as long as possible, as often as it could get it. And suddenly, up popped like and dislike functions and algorithms that deliberately elevated posts that might upset you because UI designers schooled in creating addictive software at the Stanford Persuasive Technology Lab knew that the more friction there was between you and your fellow users, the greater your number of visits and time-on-page. This would lead to more eyeballs on each ad, a higher number of click throughs, and more purchases, motivating marketers to up their social media spending.

In other words, social media executives are deliberately stoking disputes, arguments, and splits between families and friends to increase their revenue and profit margins. To do so, Tim Kendall, former Pinterest president and Facebook director of monetization and current CEO of screen time-limiting app Moment, said in a House Committee hearing:

> Tobacco companies initially just sought to make nicotine more potent. But eventually that wasn't enough to grow the business as fast as they hoped. And so they added sugar

and menthol to cigarettes so you could hold the smoke in your lungs for longer periods. At Facebook, a simple directory was engaging and kept people returning to the service. But business realities necessitated that we make the service even more engaging. To that end, we added status updates, photo tagging, and likes, which made status and reputation primary and laid the groundwork for a teenage mental health crisis.

Tobacco companies added ammonia to cigarettes to increase the speed with which nicotine traveled to the brain. Extreme, incendiary content – think shocking images, graphic videos, and headlines that incite outrage – sowed tribalism and division. And the result has been unprecedented engagement – and profits.

Facebook's ability to deliver this incendiary content to the right person, at the right time, in the exact right way; that is their ammonia. Social media preys on the most primal parts of your brain. The algorithm maximizes your attention by hitting you repeatedly with content that triggers your strongest emotions – it aims to provoke, shock, and enrage.[25]

The antidote to this particular poison is to remove the advertising incentive and ask each member to contribute a monthly or annual fee instead of duping them with the promise of "free" lifetime usage. This eliminates the need for the platform itself to kindle aggressive and combative behavior among its users. When all posts and comments are presented chronologically instead of sorted by an algorithm intended to annoy you and thereby increase the frequency and duration of your usage, all content becomes equal, leaving you to decide which posts to prioritize and which to ignore or respond to later.

Reclaiming the Promise of the Early Internet

This transition would take the power away from machine learning and put it back in your hands and those of your fellow members. It's another way that a decentralized, ad-free platform fosters a more civil and human way of interacting. The goal is to make online communication as close to face-to-face conversation as possible.

This can only happen when exploitation, monetization, and spam are eliminated entirely in a secure, safe, encrypted environment that's more like sitting down to talk freely with your friends in a pub or coffee shop than the distorted, uncivil, and rude atmosphere that Big Tech has deliberately cultivated on its money-printing platforms that masquerade as community connectors. As Jaron Lanier, futurist and author of *Ten Arguments for Deleting Your Social Media Accounts Right Now*, told *GQ* when asked about how to create more humane technology, "I think the number one priority is to not create perverse incentives that ruin quests for meaning or for happiness or for decency or betterment."[26]

That's exactly the philosophy behind decentralized social media – removing all the worst things about the internet and replacing the negatives with positives. It's impossible within the current Big Tech platforms because they lack the monetary layer needed for digital accreditation and accountability. But with a decentralized, Bitcoin-based network, you can get back to all the reasons you first loved using the web – connection, collaboration, and commerce conducted in secure, safe digital settings that closely mimic how communities work in the real world. Finally, the technology has caught up with the intentions tech ethicists outlined years ago. You no longer have to wade through the muck on other platforms because decentralized utilities offer a better alternative today.

Such an increase in accountability and civility could also be extended to email, which, as Cal Newport shares in his book *A World Without Email*, has started to become more of a hindrance than a help like it was in the halcyon "You've got mail" days. This kind of communication has long since been corrupted by the invasion of spam in our inboxes, to the degree that no amount of unsubscribing can redress the balance. But what if companies were fined 30, 50, or 100 satoshis every time they sent a spam message as with the hashcash concept mentioned earlier? This might not seem like very much, but it would quickly snowball until the amount was big enough to make it too costly to continue spamming.

CHAPTER FIVE

BUILDING A BETTER ONLINE EXPERIENCE USING BITCOIN AND THE LIGHTNING NETWORK

"While the God-fearing Man does not steal due to the threat of eternal punishment and the State-fearing Man does not steal due to the threat of legal action, the Networked Man does not steal because the encrypted Network won't let him. As networks step into their role as the new Leviathan, encryption becomes the most powerful force in the world."

BALAJI SRINIVASAN[27]

Back in chapter two, we traced the progression of the monetary system from the gold standard of the 1800s to the centralized banking of the 1900s and early 2000s to the rise of cryptocurrency since 2008. Now it's time to take this topic a little further. In the next few pages, you'll learn more about the differences between Bitcoin the asset and Bitcoin the network, the vulnerabilities of central banking, and the impact of the Lightning Network.

The Problem with Money Middlemen

One of the main challenges with a centralized financial system is that anytime there is a middleman between you and your money, they can delay, halt, or even prevent you from making payments and withdrawals. In the movie *Enemy of the State*, Will Smith's character Robert Dean uncovers a conspiracy that goes right to the top of the Federal Government. In addition to tracking his whereabouts, his adversaries start disrupting every aspect of his life. In one memorable scene, Dean tries to use his credit card to book a hotel room while he's on the run. He is shocked when his card is denied because the feds have frozen his accounts.

Such a scenario might seem like the far-fetched imagining of Hollywood scriptwriters, but in reality, what befalls Dean could happen to you or me. Your bank could decide for any number of reasons that you are suddenly persona non grata, and as a result, lock down your accounts. The IRS also has the power to go into people's private accounts and deduct money if they haven't paid their taxes. Wages can be garnished for spousal or child support. While these might all be legitimate interventions, they are also examples of how vulnerable holdings of fiat currency really are.

This doesn't just apply to banks, credit unions, and other brick-and-mortar institutions but also to apps and online payment services. For example, the non-profit National Vaccine Information Center – whose freedom-supporting mission is "Your Health. Your Family. Your Choice." – posted this website update to its supporters in late 2021:

> On December 21st past the close of business hours, NVIC was notified that PayPal would no longer process donations made by our supporters effective immediately. In essence, PayPal wants to control your choices and tell

you which non-profit charities you may and may not support. PayPal's sudden and unexplained action against our donors comes in the middle of our annual end-of-year fund raising campaign.[28]

As a result of this fiscal censorship, NVIC was forced to spend time, effort, and money setting up new payment options that allowed its financial supporters to contribute to its fundraising, much like it was forced to find alternative ways to connect with people after being deplatformed by Facebook, Twitter, and Instagram earlier in 2021. This shows how political and corporate motivations can drive a third party to put up roadblocks in the centralized financial and technology systems that not only divide buyers and sellers and creators and fans but also charities and donors. The result? More control for large corporations and less freedom for both parties whose transactions they prevented.

If NVIC had been using Bitcoin instead of PayPal, the problem wouldn't have merely been less likely to occur – it would've been impossible. Let's say I wanted to make a donation to this organization. Once I clicked "donate," NVIC would've issued an invoice. Then I'd pay this in satoshis via the Lightning Network. These funds would be deposited within seconds as the SATs flowed through the Lightning Network like a river running high. There would be no intermediary in this case, so the transaction couldn't have been slowed, delayed, or intercepted, and nor would a platform like PayPal (or another similar service like Venmo) have been able to prevent NVIC from receiving the funds.

Fiscal Freedom: Bitcoin the Asset

So how exactly does Bitcoin the asset solve the problems of the gold standard and central banking that preceded it? In his book *The Bitcoin Standard*, Saifedean Ammous wrote:

> Bitcoin's first and most important value proposition is in giving anyone in the world access to sovereign base money. Any person who owns Bitcoin achieves a degree of economic freedom which was not possible before its invention. Bitcoin holders can send large amounts of value across the planet without

having to ask for the permission of anyone. Bitcoin's value is not reliant on anything physical anywhere in the world and thus can never be completely impeded, destroyed, or confiscated by any of the physical forces of the political or criminal worlds.[29]

Part of the economic freedom that Ammous mentions derives from the fact that Bitcoin enables peer-to-peer payments via a decentralized network. Let's compare this to an old-fashioned way of transferring funds. Let's say your cousin wants to send you a birthday gift. She runs out of time to buy a present, so she pops a $50 check in the mail instead. This was issued by her bank, and when you deposit it, you must wait several days for your bank to verify with hers that she has sufficient funds to cover the amount she gave you.

Only then will the $50 be shown on the plus side of the ledger in your checking account. PayPal or Venmo are basically electronic applications that accomplish the same thing a little quicker, but as we've already seen, they still have the same main limitation: the transactions are mediated through a middleman and therefore can be disrupted or stopped on a whim. In contrast, if your cousin sent her birthday gift to you via Bitcoin, she would send the designated amount of satoshis via the Lightning Network, and you'd receive it within milliseconds, providing instant settlement.

Before we go any further, let's break down a couple of important terms. A public key is a cryptographic code that authenticates your identity as a user of a decentralized network. Just like your email address or phone number, you can share your public key without jeopardizing the security of your account. A private key is like a password that enables you to access your Bitcoin account and spend, withdraw, or transfer funds to any other user, no matter where they are in the world. The key can be a string of numbers or a 12- or 24-word phrase that you should keep secret and have a backup copy of.

With this private key, you can utilize any computer, tablet, or cell phone in the world – even if they're not yours – to conduct Bitcoin transactions safely, quickly, and securely. As Nik Bhatia wrote in *Layered Money*, "Bitcoin gives people around the world the first genuine alternative to their national currencies...Our multipolar world is looking for a monetary rebirth, and Bitcoin offers exactly that...transactions once confirmed are impossible to override, making Bitcoin the ultimate tool of financial freedom anywhere in the world."[30]

Ralph Merkle, who invented a kind of tree data structure that enhances the efficiency of Bitcoin, takes this notion further when explaining the advantages of Bitcoin over centralized banking: "It can't be changed. It can't be argued with. It can't be tampered with. It can't be corrupted. It can't be stopped. It can't even be interrupted."[31]

Another benefit of utilizing Bitcoin is that it prevents the double spending problem – i.e., the risk that currency can be spent twice. Central banking stops this from happening with fiat currency, but with some cryptocurrencies, it's difficult to verify who owns digital tokens. This can enable a digital thief to steal by duplicating a transaction to make it look legitimate, even though it isn't. In this way, the victim's account spends the same token twice. Bitcoin eliminates this possibility with the use of the blockchain (a sort of universal ledger), through which every transaction is verified.

According to a post by Team InnerQuest, "the blockchain prevents double-spending by timestamping groups of transactions and then broadcasting them to all of the nodes in the Bitcoin network. As transactions are time-stamped on the blockchain and mathematically related to the previous ones, they are irreversible and impossible to tamper with."[32]

At the Speed of Lightning

I just explained how Bitcoin gives you a greater degree of freedom, makes your financial assets tamper-proof, and prevents you from being fiscally censored when you store, send, or receive what could simply be called digital cash. But you might still be wondering how those funds your cousin sent you were available without delay. The answer is the Lightning Network. Again, Bhatia offers the best definition of this: "Lightning Network is a technological enhancement to Bitcoin that positively transforms it from a slower-moving commodity like physical gold to a currency moving at lightspeed."[33]

When precious metals were the currency of the day, certain amounts of them had to be transported between trading partners to complete transactions. This was eventually streamlined into the exchange of coins and paper money for goods and services. Next came checks and then, once electronic systems permitted it, wire transfers, credit card transactions, and payments via online platforms. For several years, sending and receiving funds via Bitcoin was like driving a car with the handbrake on – you could get to your destination, but it would be painfully slow. With the advent of the Lightning Network, your mechanic not only removed the handbrake for you but also put a Tesla drivetrain in your car, providing instant financial torque every time you push the accelerator.

Liquid Gold: Bitcoin the Network

The advantages of Bitcoin only start with its use as a financial asset. Michael Saylor, one of the pioneers of the decentralized digital transformation and chairman and CEO of MicroStrategy, outlined the unlimited growth potential that Bitcoin can provide if we think bigger and build better on top of the Bitcoin standard. Appearing on Natalie Brunell's *Coin Stories* podcast, Saylor predicted that we'll look back on Bitcoin as "digital gold without all the imperfections of gold." But its true potential could only be realized "if I plug applications into the Bitcoin network – then I can create smarter, faster, stronger gold."[34]

Saylor went on to say:

> The core principle of how you get rich or how you stay rich in the 21st century is you dematerialize something from the physical world to the digital world, and you make it smarter, faster, and stronger. If you digitize music, you create Amazon, Apple Music, and Pandora. It's smarter, faster, and stronger, and you give it to billions of people. And if you dematerialize CDs and DVDs or dematerialize concerts and Broadway shows, you have YouTube.
>
> And if you dematerialize every library and every book and every university on the planet, you have YouTube and iBooks. So the digital transformation of information and entertainment is how you provide joy and education to the world. Ergo, the digital transformation

of property is how you provide wealth to the world. It's a very simple idea. It's not only Bitcoin. It's all the applications on top of Bitcoin. All sorts of things can plug into Bitcoin, and all of them are in essence building applications and platforms on top of digital property in order to construct the 21st century cyber economy: smarter, faster, and stronger, thinking a billion times a second, moving at the speed of light....That's how people create value.[35]

Restoring Full Digital Property Rights

What Saylor is alluding to is the power of Bitcoin as a network that offers you not only the ability to conduct faster and more secure digital transactions but also the opportunity to reclaim the property rights that social media tried to take from you. This isn't the first attempt to do so. Online platforms like Substack and Patreon have gained in popularity in the past couple of years, particularly for people like Dr. Robert Malone, who was suspended from Twitter the day before a controversial December 30, 2021 interview on *The Joe Rogan Experience*.

Substack provides Malone and peers like *Pandemia* author Alex Berenson a way to share their findings with a dedicated fan base, while Patreon offers musicians, writers, and other artists the chance to connect more directly with their audiences than traditional social media and to reap the benefits of monthly patronage plans that are a kind of crowdfunding. Yet while these utilities offer some advantages over other centralized platforms, they still have one key drawback: there remains a third party mediating the posting and consuming of content, while an online payment service is responsible for processing supporters' ongoing donations.

This creates the possibility for the likes of Berenson, Malone, and others to be booted off Substack and Patreon if they are believed to have violated terms of service or committed some other arbitrary infraction that irks employees or the executive team of these third parties. In addition, a credit card or online payment service could decide to stop processing subscription payments for certain creators. The content and data of each user still belongs to the platform, and just as with Meta, Twitter, or Pinterest, the simple act of continued use is all that's needed to signal your consent with the guidelines. Another issue is that updates from these platforms are sent to subscribers via email. The day after Malone's Twitter suspension, Berenson reported that Google had started sending his updates to many people's Gmail spam folders that had previously made it to their inboxes.[36]

Such middleman meddling indicates that we need to go further toward digital liberty than the incremental steps that Substack and Patreon have empowered creators and their fans to take. Jaron Lanier proposed a radical shift in digital property rights as early as 2014, when he wrote in his book *Who Owns The Future:* "In a world of digital dignity, each individual will be the commercial owner of any data that can be measured from that person's state or behavior."[37] He then suggested that each time a member of an online community contributed something positive that elevated the group or furthered its aims, they should receive a small compensation that he termed a nanopayment. "These nanopayments will add up, and lead to a new social contract in which people are motivated to contribute to an information economy in ever more substantial ways."[38]

A Censorship-Resistant Social Connection

Decentralized networks based on Bitcoin turn Lanier's transformative proposition into reality. Not only can fans compensate creators for their work, but in turn, creators can tip fans when they contribute useful information to their groups. As with the other kinds of payments that can be made via the Lightning Network, such tips are sent securely and instantaneously. This means the sender has peace of mind in knowing that their payment will always go through, and the recipient's funds are available immediately. There's no longer a need to ask each other if they have a Venmo or PayPal account because the decentralized network renders such middlemen services obsolete.

The Lightning Network not only enables members to instantly send and receive payments but also to send messages that are totally secure and encrypted from end to end. When you do so, you will be charged a small amount, such as three satoshis. Your message will be encoded, and you will receive your three SATs back, for a net cost of zero. In this way, the platform's messaging function is like a second train endlessly running back and forth on the same Lightning Network track that delivers and receives Bitcoin transactions, both for payments in exchange for goods and services and tipping/boosting community members.

As with Bitcoin transactions, every message you send will always reach its recipient because it's secured through unbreakable encryption. And as there aren't prejudiced algorithms, distracting ads, and biased employees manipulating your data, you'll finally be able to experience genuine, unfiltered, real online community. You're no longer the product – like you've become on Facebook, Twitter, TikTok, Pinterest, and Instagram, but rather a valued and fully liberated human being who is free to speak candidly and conduct commerce directly.

Saylor told Brunell that "Thomas Watson built on the mainframe platform, Bill Gates built on the PC platform, and Jeff Bezos built on the internet platform. And then Apple came back and revitalized the mobile platform. Now you've got the next platform: digital property is a platform."[39] Decentralized utilities go one step further by building on the Bitcoin standard, enabling you to achieve both digital sovereignty and fiscal freedom in a censorship-proof environment. The goal? To give you greater control over your financial assets, more freedom to express yourself, and the ability to connect with creators, family, and friends around the world in a rich, authentic way that mirrors how you'd interact with them in person. How do we get there? Bitcoin the network.

CHAPTER SIX

UNAPOLOGETIC FREEDOM

"Movements are what take five or ten percent of people and make them decisive – because in a world where apathy rules, five or ten percent is an enormous number."

BILL MCKIBBEN[40]

In previous chapters, we've examined the current state of censorship, the evolution of the monetary system from the gold standard to central banking to Bitcoin as we transitioned from a centralized century to a decentralized one, and the differences between platform and peer governance. Then we looked at how Bitcoin the asset is like liquid gold that brings fiscal freedom, how the Lightning Network accelerates this transfer of digital wealth, and why Bitcoin the network holds the promise of a new era. Now it's time to land the plane and let you know how you can take back your freedom of and after speech, engage in censorship-proof community and commerce, and utilize a decentralized network to claim full sovereignty of your digital identity.

You are Not a Product

In his book *Freedom*, Hong Kong's leading pro-democracy activist and Nobel Peace Prize nominee Nathan Law wrote, "When governments control access to information and are able to define the narrative and dictate what we know, we lose more than our freedoms. We lose the ability to see the world for what it is. We lose our humanity."[41] The same is true when Big Pharma, Big Tech, and other corporate interests wield what C.S. Lewis once called "hideous strength" to advance agendas of profit and control and, in doing so, limit your freedom by censoring you or removing you from the public discourse.[42]

As you've seen earlier in this book, the solution is to bypass the centralized banking and social networks that such nefarious forces control and, to revisit a previous analogy, live in a house you've built on your own land, instead of being a tenant at risk of eviction on the property of a corrupt landlord. In making this move, you'll decide what gets said and done in your own online community, and – if you're using an ad-free, censorship-resistant network – the platform cannot punish you for saying or doing what they consider to be the wrong thing (although you will be held accountable for speaking civilly to fellow community members by your peers).

It would be nice to think that government regulators or the conscience of Big Tech billionaires would lead them to either break up their centralized monopolies or change their business model so that people were prioritized over profits. But as Sheera Frenkel and Cecilia Kang state in their book *An Ugly Truth*, "The algorithm that serves as Facebook's beating heart is too powerful and too lucrative. And the platform is built upon a fundamental, possibly irreconcilable dichotomy: its purported mission to advance society by connecting people while also profiting off them."[43]

This isn't just true of Facebook but also of Twitter, Instagram, LinkedIn, TikTok, Pinterest, and whatever new centralized networks might have sprung up since I wrote these words. Which is why Frenkel and Kang believe that the trend of Zuckerberg and co. making "massive gains [that] have repeatedly come at the expense of consumer privacy and safety and the integrity of democratic systems" will continue.[44]

As such, the only way for you to find freedom from censorship, manipulation for profit, and the biggest social engineering experiment in human history is to step away from central platforms and embrace decentralized ones that can prove they protect your liberty, individuality, and privacy. Several years ago, Google announced its new mantra: Don't be evil. And yet as we've seen with ad-driven distortion of search results and politically-motivated suppression of free speech, the company's desire to satisfy its advertisers and the biases of its employees can be brought to bear in a damaging way on an unsuspecting public. And according to former employees in an article published by *NPR*, Google violated its own "don't be evil" pledge by being involved in "controversial projects."[45]

A better slogan would be "Can't be evil." This is because messages between members and Bitcoin transactions alike are encrypted from end to end, meaning that staff can't even see the data. Unlike centralized networks, decentralized ones also cannot store, report on, or sell your information because it's impossible when using Bitcoin as a network as well as an asset. This way, you can trust that every time you boost (aka tip) a fellow community member or send them SATs, your funds will go through and be received in mere milliseconds. Similarly, when you send a message, it won't be read, blocked, or altered, as can happen with the Federal Government, traditional social networks, and even messaging, phone, and email services.

Masters and Slaves of Money

Let's take one final look at the differences between storing and moving your money through Bitcoin versus doing so via the central banking system. In a blog post, freedom maximalist and ex-hedge fund manager Robert Breedlove wrote, "Money is a tool for trading human time. Central banks, the modern-era masters of money, wield this tool as a weapon to steal time and inflict wealth inequality. History shows us that the corruption of monetary systems leads to moral decay, social collapse, and slavery."[46]

In contrast, he asserted:

> Bitcoin is a rebellion against the most powerful bastion of socialism in the free world: central banking. It is a peaceful revolution involving the permanent disarmament of tyrants that weaponize money to confiscate wealth. Bitcoin is a weapon of peace used for the assassination of time-theft. An alchemical archetype, it is an antidote to state corruption and social moral affliction.

> As a purely honest free market money, Bitcoin is an irrepressible truth; an expression of pure monetary capitalism and a modern-day declaration of independence for fiat-slaves worldwide. Bitcoin is money without masters: a system governed by rules instead of rulers. By awakening the world from the nightmare of financial slavery, Bitcoin is a dream of freedom coming true.[47]

It's this dream that inspired me to take a stand, team up with the best partners, and create a company based on the Bitcoin network. You as an individual have an opportunity to do the same, just by owning Bitcoin. It's the last sovereign asset available in the world. As we've already seen, if you deposit money in a central bank, it's no longer yours. If you have a mortgage, your home belongs to someone else. Even the electricity, water, and other utilities you use are owned by third parties. But Bitcoin can give you true ownership of digital property and independence forever.

Sometimes the combined power of centralized social and monetary networks can seem scary and unshakeable. But while these entities will continue trying to claw back our march toward liberty as we progress in this decentralized century, you have the option to choose something better every time you open your wallet or your browser. Instead of Googling, use Brave or DuckDuckGo. Utilize ProtonMail or Tutanota to send encrypted emails. Rather than refreshing your Facebook, Twitter, or Instagram feed, talk freely and openly on a decentralized, censorship-proof social network. Store, send, and receive Bitcoins instead of using your credit card and checking account. And question what other changes you can make to give your life more liberty and freedom from censorship and overreach.

If we all make and repeat such choices daily, we can shake off the shackles of the centralized networks and make lasting, positive change for ourselves, our children, and our grandchildren. As JP wrote in the foreword, we have reached a tipping point for our very civilization, and standing idly by is no longer an option. We're either going to continue perpetuating control, censorship, and suppression by giving our attention, time, and energy to central platforms that wish only to exploit us for their own gain, or we'll elect to choose liberty, unstoppable speech, and freedom by embracing Bitcoin via decentralized utilities.

Speaking on *Palisades Gold Radio*, entrepreneur and cryptocurrency expert Mark Moss stated: "I know a lot of this stuff is scary. But I'd like to encourage people that the human drive for freedom always wins. I'm extremely hopeful for the future. We're not trees – we can get up and move. And we're not just sheep. We have a voice, and the people ultimately have the power. Don't be a sheep, don't be a pushover."[48]

Not Your Keys, Not Your Coin

Throughout history, the possession of keys has symbolized ownership of property and unfettered access to it. Certain cultures even have key-related ceremonies, such as the British Army's parade that marks the handover of the keys to the Tower of London when a new unit starts their guard duty of this world-famous building.

The same concept of property ownership is true of Bitcoin, which is actually even better because unlike a physical building, digital property in Bitcoin will only ever be yours. As I explained in chapter five, when you decide to buy Bitcoin for the first time, you will be given two different keys. The first is a public key that anyone can view without the security of your digital wallet being compromised, while the second is a private key that you should keep to yourself and back up so you don't lose access to your funds. The latter gives you unseizable property rights on the Bitcoin blockchain instead of placing your money with a centralized bank that is subject to corporate and governmental actions that can limit or remove access to it.

Using Bitcoin is like having your own personal bank trading in liquid gold that can be broken into little pieces (satoshis) and sent around the world at the speed of light. And unlike with traditional transactions, you're protected by smart contracts. With Bitcoin the asset, money is unhackable, unstealable, and unstoppable. This newfound ability to own secure digital property offers you the kind of financial freedom you've never had, and nobody can take it away from you. Encryption creates invulnerable digital property rights, and as you control your keys, you control your freedom. If you're new to cryptocurrency and need further guidance, reading *Layered Money* by Nik Bhatia or listening to Mark Moss's podcast can help you get started.

UNAPOLOGETIC FREEDOM

We Won't Censor You vs. We Can't Censor You

When you choose to use a decentralized network, the same fiscal liberties that you'll enjoy with Bitcoin the asset also apply to your online interactions via Bitcoin the network. As I wrote earlier, we figured out a way to piggyback the transfer of data on the same rails that send digital money back and forth. As a result, whatever you say is truly private because of beginning-to-end encryption. Services like WhatsApp claim that they won't censor you, even though many tech experts believe that information is being sliced, diced, and sold much like it is on old-style social networks owned by the same parent companies.

On a decentralized network, the platform cannot censor you because your data cannot be collected or stored, let alone peddled for profit. Centralized networks implore you to trust them, even though they prove to be untrustworthy time and time again. We urge you to verify instead: do your own research and then decide which platforms offer you the most privacy, security, and freedom.

You should also decide what kind of online environment you want yourself and your family to use. Do you like being shown ad after ad, knowing that your information is being sold, and being herded toward information that upsets or angers you? Or do you wish for something better? If it's the latter, then you'd probably prefer a decentralized platform where there are no hyper-targeted ads, no increasingly-polarizing "recommended content," and no threat of surveillance or arbitrary censorship. Somewhere that rules are enforced by technology and the accountability of peer governance — not Big Tech companies' platform governance that's driven by their greed and control.

The latest Bitcoin-based technology means that for the first time, you can speak as candidly online as you would in your best friend's living room without the risk of being silenced. This is how we're removing the barrier between in-person and online community and freeing you to be a person, not a product.

Ever since centralized social platforms settled on their ad-driven monetization model, you became part of the attention economy. The goal was to exploit rather than connect you, and you were turned into a servant of the technology. In contrast, we believe that technology should serve you. That's why communities on a decentralized network are fully independent micro-economies that unlock opportunities for anyone to create wealth using Bitcoin the asset. By combining this with all the advantages of Bitcoin the network, a decentralized utility makes it easier for artists to pursue what they love and share their passion with the world, while fans have direct access to content, those who create it, and friends and family everywhere.

You deserve to have a place to speak your truth and to do so safely, securely, and civilly, without the fear of being censored or punished for violating arbitrary terms of service. You should also be able to accumulate, send, and receive digital wealth without central banking or governments interfering. We believe that Bitcoin provides a way to clean up cyberspace, remove troublesome middlemen, and eliminate censorship. With it, you can restore your digital property rights, declare full sovereignty, and express unapologetic freedom. This is the path to a better future, and that future starts now.

NOTES

NOTES

ACKNOWLEDGEMENTS

I want to thank Phil White for taking a chance on me and agreeing to help write this manuscript. Thanks to my family – Mom (Jouliet), Dad (Mike), and Brother (Jonathan) for always being supportive through my toughest times. My cousin Reza, his wife Isa, and their kids, Lou Rose and Lucas, for making me laugh and remember that the best times in life are those moments spent with family. Thank you to my team that made this all possible and JP Sears for saying yes to joining me on this amazing journey.

I'm also grateful to all our Investors & Advisors. Aubrey, Tony, and Mark, you inspire me to be a better executive each day. Thank you to every Zion member for continuing to support our mission to build a decentralized future. Finally, I would like to thank Satoshi Nakamoto.

ABOUT THE AUTHOR

Justin Rezvani is a first-generation American entrepreneur, Ironman triathlete, explorer, and member of the 2017 *Forbes* 30 Under 30 list. Rezvani received his Bachelor of Science in business administration, marketing management, and advertising from California State Polytechnic University, Pomona, where he was named student of the year in both 2010 and 2011. He began his professional career at The Walt Disney Company. In 2013, at age 25, Rezvani founded theAmplify, a platform that creates large-scale native advertising campaigns for brands on platforms like Instagram, Facebook, Snapchat, and YouTube. Building one of the world's first influencer platforms with no outside investment or funding, Rezvani bootstrapped theAmplify and the company was cash flow positive within six weeks.

The impressive executive team that Rezvani assembled allowed the company to successfully expand to become a profitable, eight-figure business in two years, with clientele across all business verticals. Rezvani was an internal marketing advisor to Ford, Lionsgate, Pepsi, Unilever, COTY, and Campbell's during his time as CEO, crafting some of the most iconic influencer campaigns in the world. In April 2016, theAmplify was acquired by You & Mr. Jones.

In early 2018, Rezvani left his role at theAmplify, and in 2020, he began developing Zion, the social network built on Bitcoin. It is the next generation of social media that facilitates the free and open flow of content and payments between artists and their audiences. Free of censorship, Zion is a decentralized network for creative people to pursue what they love and share their passion. It makes the world better with trusted, censorship-proof, ad-free social media.

Rezvani has been a notable keynote speaker at global marketing events such as Cannes Lions festival, CES, The CMO Club, StreamCon, Playlist

ABOUT THE AUTHOR

Live, and the AAF and has been featured in articles in *The Wall Street Journal*, *Inc.*, *Adweek*, *Ad Age*, *Business Insider*, *LA Times*, *Los Angeles Business Journal*, and more. Follow him at @justinrezvani on all major platforms.

END NOTES

1. Gyan Nagpal, *Talent Economics*, (London: Kogan Page, 2015), 40.
2. Nik Bhatia, *Layered Money*, 2021, 64-65.
3. "The Battle of Bretton Woods," *Council on Foreign Relations* summarizing Benn Steil, *The Battle of Bretton Woods* (Princeton, NJ: Princeton University Press, 2013), available online at https://www.cfr.org/book/battle-bretton-woods.
4. "Alphabet Agencies," Wikipedia, available online at https://en.wikipedia.org/wiki/Alphabet_agencies.
5. Bhatia, *Layered Money*, 97.
6. Ibid, 112.
7. Ibid, 113.
8. Balaji Srinivasan, Twitter, September 4, 2021, accessible online at https://twitter.com/balajis/status/1434209685066838016.
9. Balaji Srinivasan, Twitter, February 24, 2018, available online at https://twitter.com/balajis/status/967552050761363456.
10. "Facebook: Transparency and Use of Consumer Data," April 11, 2018, House of Representatives, Committee on Energy and Commerce, available online at https://docs.house.gov/meetings/IF/IF00/20180411/108090/HHRG-115-IF00-Transcript-20180411.pdf.
11. Roger McNamee, *Zucked* (New York: Penguin, 2019), 129.

12 Fiona Godlee and Kamran Abbasi, "Covid-19: Researcher Blows the Whistle on Data Integrity Issues in Pfizer's Vaccine Trial. Rapid Response: Open letter from The BMJ to Mark Zuckerberg," *The BMJ*, December 17, 2021 update to a paper first published on November 2, 2021, available online at https://www.bmj.com/content/375/bmj.n2635/rr-80.

13 "For Technologists," Center for Humane Technology, available online at https://www.humanetech.com/technologists.

14 Kenny Kane, "Want to Make Your Athletes More Fit, Healthy, and Well? Your Online Habits Need to Change," TrainHeroic, 2019, available online at https://www.trainheroic.com/blog/want-to-make-your-athletes-fit-healthy-and-well.

15 James Williams, *Stand Out of Our Light* (Cambridge, England: Cambridge University Press, 2018), 7-8.

16 Tim Wu, *The Attention Merchants* (Knopf: New York, 2017), 344.

17 Balaji S. Srinivasan, "The Network State," Foresight Institute, available online at https://foresight.org/salon/balaji-s-srinivasan-the-network-state.

18 Silke Helfrich and David Bollier, "Peer Governance Through Commoning," Resilience.org, February 26, 2020, available online at https://www.resilience.org/stories/2020-02-26/peer-governance-through-commoning.

19 Srinivasan, "The Network State," Foresight Institute.

20 Balaji S. Srinivasan, "The Network State," November 11, 2021, available online at https://1729.com/the-network-state.

21 Andrea Weckerle, *Civility in the Digital Age* (London: Que Publishing, 2013), 4.

22 Jennifer Murtell, "Empathy and Technology: Shaping Our Connection in the Future," AMA, September 28, 2020, available online at https://www.ama.org/marketing-news/empathy-and-technology-shaping-our-connection-in-the-future.

23 Paul Itoi, "Freedom from Free," Sphinx, 2021, available online at https://blog.sphinx.chat/2021/02/12/freedom-from-free.

END NOTES

24. Ian Tucker, "Sherry Turkle: 'The Pandemic Has Shown Us that People Need Relationships'," *The Guardian*, March 21, 2021, available online at https://www.theguardian.com/science/2021/mar/21/sherry-turkle-the-pandemic-has-shown-us-that-people-need-relationships.

25. House Committee on Energy and Commerce Testimony of Tim Kendall, September 24, 2020, available online at https://www.congress.gov/116/meeting/house/111041/witnesses/HHRG-116-IF17-Wstate-KendallT-20200924.pdf.

26. Zach Baron, "The Conscience of Silicon Valley," *GQ*, August 24, 2020, available online at https://www.gq.com/story/jaron-lanier-tech-oracle-profile.

27. Srinivasan "The Network State," Foresight Institute.

28. "PayPal Stops Processing NVIC Donations," National Vaccine Information Center, December 24th, 2021, available online at https://www.nvic.org/nvic-vaccine-news/december-2021/paypal-stops-nvic-donations.aspx.

29. Saifedean Ammous, *The Bitcoin Standard* (Hoboken, New Jersey: Wiley, 2018), 200.

30. Bhatia, *Layered Money*, 149, 152.

31. Ammous, *The Bitcoin Standard*, 175.

32. Team InnerQuest, "How Does a Blockchain Prevent Double-Spending of Bitcoins?" *Medium*, August 25, 2018, available online at https://medium.com/innerquest-online/how-does-a-blockchain-prevent-double-spending-of-bitcoins-fa0ecf9849f7.

33. Bhatia, *Layered Money*, 126.

34. "Michael Saylor In-Depth: Bitcoin is Fire or Electricity," *Coin Stories*, July 15, 2021, available online at https://www.youtube.com/watch?v=_Ae134VhSUA&ab_channel=NatalieBrunell.

35. Ibid.

END NOTES

36 Alex Berenson, "Good News and Bad News on the Stack," Unreported Truths, December 30, 2021, available online at https://alexberenson.substack.com/p/good-news-and-bad-news-on-the-stack/comments.

37 Jaron Lanier, *Who Owns the Future?* (New York: Simon & Schuster, 2014), 20.

38 Ibid.

39 "Michael Saylor In-Depth: Bitcoin is Fire or Electricity," *Coin Stories*.

40 Paul Hawken, *Drawdown* (New York, Penguin, 2017), 216.

41 Nathan Law, *Freedom* (New York: The Experiment, 2021), 49.

42 C.S. Lewis, *That Hideous Strength* (New York: Scribner, 2003), 300.

43 Sheera Frenkel and Cecilia Kang, *An Ugly Truth* (New York: HarperCollins, 2021), 300.

44 Ibid.

45 Bobby Allyn, "Ex-Google Workers Sue Company, Saying it Betrayed 'Don't Be Evil' Motto," *NPR*, November 29, 2021, available online at https://www.npr.org/2021/11/29/1059821677/google-dont-be-evil-lawsuit.

46 Robert Breedlove, "Masters and Slaves of Money," July 5, 2020, available online at https://breedlove22.medium.com/masters-and-slaves-of-money-255ecc93404f.

47 Ibid.

48 "Mark Moss: How to Prepare for the Great Reset," *Palisades Gold Radio*, October 20, 2021, available online at https://www.youtube.com/watch?v=Y8MdJyc_M5w&ab_channel=PalisadesGoldRadio.

Made in the USA
Middletown, DE
21 March 2023